Ductile Fracture and Ductility

With Applications to Metalworking

BRADLEY DODD

Department of Engineering
University of Reading
Reading, UK

YILONG BAI

Institute of Mechanics
Chinese Academy of Sciences
Beijing, People's Republic of China

1987

ACADEMIC PRESS
Harcourt Brace Jovanovich, Publishers
London   Orlando   San Diego   New York   Austin
Boston   Sydney   Tokyo   Toronto

ACADEMIC PRESS INC. (LONDON) LTD.
24/28 Oval Road, London NW1 7DX

United States Edition Published by
ACADEMIC PRESS INC.
Orlando, Florida 32887

British Library Cataloguing in Publication Data
Dodd, Bradley
Ductile fracture ductility
1. Metals—Fracture   2. Metals—Ductility
I. Title   II. Bai, Yilong
620.1′66   TA460

ISBN 0-12-219125-0

Typeset by Eta Services (Typesetters) Ltd, Beccles, Suffolk
and printed in Great Britain by
St Edmundsbury Press Ltd, Bury St Edmunds, Suffolk

Preface

To obtain a fundamental understanding of ductile fracture and ductility it is necessary that engineers and metallurgists work together closely. The reasons for this are twofold. First, it appears to be impossible for either macroscopic mechanics or metallurgical micromechanisms of fracture to achieve this aim separately. Secondly, advances in instrumentation and theories have helped to make the subject interdisciplinary.

It is the interdisciplinary nature of ductile fracture that originally brought the authors together from their different backgrounds. Moreover, we did not allow the large distance between our respective Institutions to become an obstacle to the completion of the work.

There is an extensive literature on ductile fracture spread over numerous journals, so inevitably some decision had to be made about the exclusion of certain areas of literature. Therefore we decided to confine the description in the book to ductile fractures that do not occur at elevated temperatures. Thus modes of creep deformation are not described.

We decided to take a particular unified approach to the subject rather than to produce just a "data bank". Specifically, we have concentrated a large portion of the book on submillimetre domains and their interactions. Thus void nucleation and growth and shear bands and their interactions often play a significant role in ductile fracture. We accept that this approach is highly personalized, but hope that it reflects the relative importance of the various mechanisms.

Although it has been necessary to use mathematics to elaborate the theories of ductile fracture, we have made a conscious effort to give the mathematics used a sound physical basis. It is clear to us that detailed models of ductile fracture are required, but these must be based on firm physical foundations. Thus tensor analysis is used, but in a simplified manner. It is hoped that the material presented in this way will be accepted favourably by our readers.

This book will be of interest to engineers and metallurgists both in industry and educational establishments. Specifically, senior undergraduate and postgraduate students will find this book of interest, as will researchers working in the area of fracture.

We are indebted to the Royal Society and the Chinese Academy of Sciences for arranging for one of us to visit the U.K. It was during this visit that the writing programme for this book was initiated. We are also indebted to Dr Elizabeth Frankland-Moore, O.B.E., of the Sino–British Fellowship Trust and the Henry Lester Trust for financial help that facilitated a further visit to the U.K. by Y.L.Bai during which the manuscript for the book was completed. We are also very grateful to Professors W. Johnson, F.R.S., A. G. Atkins and P. D. Dunn for helpful discussions during the preparation of the manuscript. Acknowledgements are due to Professor C. M. Cheng and the Institute of Mechanics, Beijing, for the facilities offered to one of the authors.

Drs P. Hartley and D. V. Wilson read the entire manuscript, and their discussions have been particularly helpful.

We should like to thank Mr H. Kobayashi for his excellent drawings, which were used where photographs were not obtainable.

Finally, above all, this book would not have been possible had it not been for the help and contributions from our friends and colleagues, knowingly and unknowingly.

Authors' royalties for this book will be sent in equal proportion to the Sino–British Fellowship Trust and the Band Aid Trust.

July 1986   *Bradley Dodd*
                                                                          Yilong Bai

Contents

3 Effects of Hydrostatic Stress, Temperature and Strain Rate on Ductility   41

4 Void Nucleation, Growth and Coalescence   73

5 Theories of Ductile Fracture I: Void Dynamics   97

6 Adiabatic Shear   123

7 Theories of Ductile Fracture II: Localization of Plastic Deformation   147

To Ruth and Fujiu Ke

Nomenclature

A	cross-sectional area
a	crack length
B	thickness of specimen
b	Burgers vector
COD	crack opening displacement
c	wave velocity
c_v	specific heat
d	grain size; diameter
E	Young's modulus
e	elongation; engineering strain
F	force
G	shear modulus
K	stress intensity factor; bulk modulus
K_c	fracture toughness
k	shear yield stress
L	load, length
M	torque
n	strain hardening exponent
n_i	normal vector
p	hydrostatic pressure
R	reduction in area; radius
r	reduction in height; radius; strain ratio
T	temperature
t	thickness
U	energy
u	displacement

V	volume
v_i	velocity
W	width; strain energy density
α	strain ratio
γ	shear strain; surface energy
$\dot{\gamma}$	shear-strain rate
δ	crack opening displacement; band width
ε_{ij}	strain tensor
ε_f	critical strain; fracture strain
$\bar{\varepsilon}$	effective strain
$\dot{\varepsilon}$	strain rate
θ	angle
κ	thermal diffusivity
Λ	spacing
λ	thermal conductivity; length
$\dot{\lambda}$	non-negative multiplier
μ	modulus
ρ	radius; density
Σ	macroscopic stress
σ_{ij}	stress tensor
σ_f	fracture stress
σ_m	mean normal stress
σ_y	yield stress
$\bar{\sigma}$	effective stress
$\dot{\sigma}$	stress rate
τ	shear stress
ϕ	angle
Ω	atomic volume

Other symbols used are defined in the text.

1

Introduction to Ductile Fracture Mechanisms

1.1 Features of ductile fractures

The mechanics and mechanisms of ductile fracture in metals are of significance for both engineers and metallurgists in designing against plastic collapse and fracture of structures and workpieces. The avoidance of ductile fracture is of importance in metal-forming processes, such as forging, extrusion and drawing. Since the 1950s, several helpful concepts have been proposed, such as fracture toughness K_{IC}, the J-integral and the crack opening displacement (COD). These concepts have been used successfully in studies of some fracture processes, particularly in design. However, the fracture tests based on these concepts exhibit, to a greater or lesser extent, size effects. It is worth noting that plastic energy is measured as an energy per unit volume, whereas surface energy is measured, naturally, as an energy per unit area. Because both elastic strain energy and plastic energy are involved in ductile fractures, perhaps the only way to combine the two energy functions is to assume an internal size-effect for the plastic work, as did Orowan (1955). This implies that it may be extremely difficult, or even impossible, to identify a single material constant for the combined elastic and plastic deformations that occur in ductile fractures. For purely elastic systems it was shown by Griffith (1921) that it is possible to derive a brittle fracture criterion in terms of the surface energy of the crack.

It is not rare in engineering design against fracture that large plastic deformations need to be taken into account. Metal-forming processes (both sheet and bulk forming), ordnance applications and the deformation ahead of a crack tip are all examples. Thus the problems raised here are of significance both industrially and academically. Confining attention to tensile loading,

1

BROAD CLASSES OF FRACTURE MECHANISM

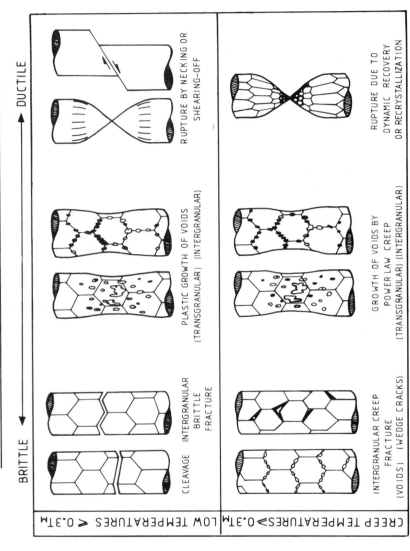

Fig. 1.1 Classification of fracture mechanisms. (After M. F. Ashby, *Prog. Mat. Sci.*; *Chalmers Anniversary Volume*, pp. 1–25 (1981).)

Ashby (1981) has listed three types of fracture (see Fig. 1.1): cleavage, nucleation and growth of voids, and rupture.

1.2 Cleavage

Cleavage is usually considered to be a form of brittle fracture that is characterized by no apparent deformation in the vicinity of the crack. Although the cleavage crack propagates in a brittle manner, the cracks are nucleated by three different mechanisms. Accordingly, cleavage is subdivided into three types.

For cleavage 1, pre-existing cracks are large enough to allow cracks to propagate at a stress lower than the yield stress at which activation of slip systems occurs and plasticity is very limited.

Solids with or without pre-existing cracks can fracture by cleavage 2. In this regime cleavage nuclei are initiated by slip bands or twins. Observations suggest that cleavage nuclei are initiated by yielding before fracture. In this case a shear band or twin can be blocked by a grain boundary or second phase particle. Thus it is reasonable to suppose that the nucleated cracks have a length proportional to the grain size d. The fracture stress σ_f can be defined as

$$\sigma_f = \left(\frac{EG}{\pi d}\right)^{1/2},\tag{1.1}$$

where E is Young's modulus and G is the toughness. The fracture stress depends on the grain size in two ways (Fig. 1.2). If the yield stress σ_y is greater

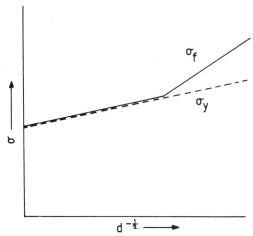

Fig. 1.2 Schematic variation of yield and fracture stresses with grain size.

than σ_f then fracture occurs at σ_y, as shown to the left of Fig. 1.2. When σ_f > σ_y (to the right-hand side of Fig. 1.2) cleavage nuclei are initiated at stress σ_y, but cannot propagate until σ_f is reached. Therefore cleavage 2 is different from cleavage 1 because of the localized plastic zone at the crack tip.

Cleavage 3 is distinguished from the other two regimes of cleavage by more extensive plastic flow (strains of about 1–10%) because of the low yield stress. General plasticity can nucleate cracks at grain boundaries, allow them to grow stably and then, at some critical length, propagate as cleavage cracks.

Like ductile fracture, cleavage is usually transgranular. However, impurities or imperfections at grain boundaries can produce brittle intergranular fracture. Ashby names these fractures brittle intergranular fracture (BIF). Based on different modes of nucleation similar to those in cleavage, three similar subgroups of BIF are possible.

1.3 Voids

A typical mode of ductile fracture is by void growth. The process consists of three sequential stages: void nucleation, void growth and void coalescence (Fig. 1.3).

Stress concentrations at hard particles can lead to local plasticity, which in turn can lead to inclusion cracking or inclusion-matrix separation. The nucleation strain, beyond which void nucleation begins, is found to occur at

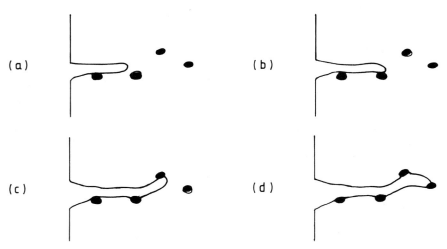

Fig. 1.3 Nucleation, growth and coalescence of voids: (a) void nucleation by particle-matrix decohesion; (b) crack extension by joining with isolated voids; (c) and (d) continued extension with more plastic deformation.

strains between almost zero and unity. The nucleation of a void at a particle is strongly dependent upon how the particle is bonded to the matrix. If the bonding is fairly weak (e.g. manganese sulphide in steel) then voids can nucleate at low stresses and small strains. However, if the particle is bonded to the matrix strongly (e.g. carbides in steel) then large strains precede nucleation.

After nucleation, voids grow under continued loading. Undoubtedly an initially spherical void would become ellipsoidal with plastic deformation. However, the physics of void growth is complex. Obviously plasticity can enhance void growth. This happens often in engineering and will be discussed in detail in this book. There are other mechanisms for void growth, for example creep. The strain-rate dependence involved in creep can postpone void nucleation to larger strains. Also, this dependence stabilizes plastic flow until larger strains, postponing void coalescence.

Ductile fracture is normally transgranular. However, if the volume fraction of voids on grain boundaries is large then void growth can take place in an intergranular manner. Under these circumstances, ductile fracture is termed fibrous. Intergranular ductile fracture occurs in low-stress long-time creep tests, whereas at high temperatures diffusion can cause void growth. Void coalescence is the final stage in void-controlled ductile fracture. It is usually assumed that plasticity is localized between the voids. This localized deformation leads to final coalescence of voids and complete material separation. These three sequential steps for fracture by voids constitute a major feature of ductile fracture.

1.4 Rupture

Rupture, in its general sense like necking and shearing-off, is a further type of ductile fracture which is associated with plastic instability. In simple tension, plastic deformation proceeds uniformly along the entire gauge length of the specimen until a critical condition is attained beyond which deformation becomes localized, forming a neck or a shear band. With further extension, the localized plastic zone will become more pronounced and eventually the cross-section will become zero. This instability-induced rupture is closely related to the geometrical configuration of the specimen, strain hardening, strain-rate sensitivity and the softening of the material.

A prerequisite for pure rupture is the suppression of void nucleation and growth. Situations where rupture is possible are materials of high purity and plastic deformation occurring simultaneously with dynamic recrystallization (Fig. 1.1). In general, strain and strain-rate hardening oppose early local plastic instability and therefore enhance the possibility of rupture. On the

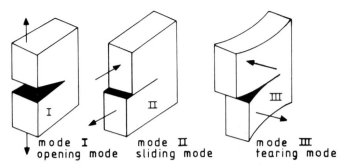

mode **I**   mode **II**   mode **III**
opening mode   sliding mode   tearing mode

Fig. 1.4   The three modes of fracture.

other hand, the greater the geometrical constraint, the greater is the possibility of void nucleation and growth.

It is clear that plastic instability not only plays a crucial role in rupture, but also in void-induced ductile fracture. Local shearing can lead to void sheeting, for example. In addition, plastic instability can influence the nucleation and growth of voids. It is now generally recognized that plastic instability plays an essential role in a variety of ductile fractures.

1.5   Modes of fracture

There are three different modes in which a crack can be stressed; these are shown in Fig. 1.4. Mode I is the so-called opening mode, and cracks are stressed in this manner in measures of fracture toughness in the compact-tension and double-notched bend specimens. For each mode it is possible to derive the stress and deformation fields directly, as shown by Knott (1973).

1.6   Linear elastic fracture mechanics (LEFM)

For brittle materials LEFM can be used to describe fracture. The origins of fracture mechanics lie in Griffith's energy criterion for the catastrophic propagation of pre-existing cracks in a material. According to Griffith's criterion, a crack will propagate in a brittle material when the decrease in elastic strain energy is at least equal to the energy required to create the new crack surface. Using this criterion, it is possible to determine the tensile stress necessary for crack propagation. For a thin sheet containing an edge crack of length a (or equivalently an internal crack of length $2a$) (see Fig. 1.5) subjected to an axial stress σ, the decrease in the elastic strain energy per unit

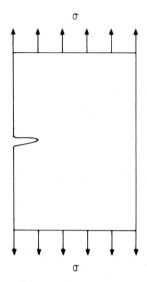

σ

Fig. 1.5   A thin sheet with an edge crack subjected to an axial stress.

thickness due to the crack is

$$U_e = -\pi a^2 \sigma^2 / E,$$

where E is Young's modulus. The increase in surface energy due to the crack is $U_s = 4a\gamma_s$, where γ_s is the surface energy per unit area. We therefore have

$$U = U_e + U_s. \tag{1.2}$$

Now for equilibrium $dU/da = 0$, which leads to

$$\sigma = \left(\frac{2E\gamma_s}{\pi a}\right)^{1/2}. \tag{1.3}$$

Orowan (1950) suggested that localized plasticity at the crack tip could be taken into account with an additional term γ_p, which is a measure of the plastic work required to extend the crack; thus

$$\sigma = \left\{\frac{2E(\gamma_s + \gamma_p)}{\pi a}\right\}^{1/2}. \tag{1.4}$$

The application of LEFM provides a full description of the stress and strain fields in the vicinity of the crack. From linear elastic theory the stresses in the region of the crack tip are given by

$$\sigma_{ij} = \frac{K}{(2\pi r)^{1/2}} \, f_{ij}(\theta), \tag{1.5}$$

where r and θ are the cylindrical polar coordinates of a point with respect to the crack tip and K is called the stress intensity factor. In general, the stress intensity factor is given by

$$K = \sigma(\pi a)^{1/2} f\left(\frac{a}{W}\right), \tag{1.6}$$

where $f(a/W)$ is a dimensionless parameter whose value depends on the geometry of the specimen. In the case of the compact-tension specimen (shown in Fig. 1.6) the critical stress intensity factor in mode I loading, K_{IC}, is measured. To take account of crack-tip plasticity and limited ductility of some materials, the plane-stress plastic-zone radius given by $r_y = (2\pi)^{-1}(K_{\mathrm{I}}/\sigma_y)^2$ should be, empirically, no larger than one-tenth of the specimen thickness. To ensure plane-strain conditions, the following dimensional restrictions apply: $a = B = \frac{1}{2}W > 2.5(K_{\mathrm{IC}}/\sigma_y)^2$, where a is the crack length, B the thickness and W the width of the specimen.

Table 1.1 lists the limiting size requirements for a range of materials. It is important to note that a testpiece of about 600 mm in thickness would be necessary for ductile A533B-1 pressure-vessel steel; an unrealistic test!

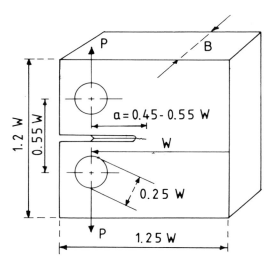

Fig. 1.6 Compact-tension specimen.

Table 1.1

Approximate limiting size requirements for characterization by linear elastic fracture mechanics in different materials (after R. O. Ritchie, *Trans. ASME*, H: *J. Engng Mat. Technol.* **105**, 1 (1983))

Material	σ_y (MPa)	K_{IC} (MPa m$^{1/2}$)	r_y (μm)	Limiting size (mm)	
4340, 200 °C temper	1700	60	200	3	($\sim$0.1 in.)
Maraging steel	1450	110	920	14	($\sim$0.5 in.)
A533B-1	500	245	4×10^4	600	($\sim$2 ft)
7075-T651	515	28	470	7	($\sim$0.3 in.)
2024-T351	370	35	1420	22	($\sim$1 in.)
Ti-6Al-4V	850	120	3170	50	($\sim$2 in.)
Tungsten carbide	900	10	20	0.3	($\sim$3 mils)
Polycarbonate	70	3	290	5	($\sim$0.2 in.)

Figure 1.7 shows the approximate ranges of application of LEFM and elasto-plastic fracture mechanics.

1.7 Elasto-plastic fracture mechanics (EPFM)

For elasto-plastic conditions, for which crack-tip plasticity cannot be ignored, the *J*-integral and the crack opening displacement (COD) have been used.

The *J*-integral, developed by Rice (1968), is given by

$$J = \int_F W \, dy - T_i \frac{\partial u_i}{\partial x} \, ds \tag{1.7}$$

(see Fig. 1.8), where T$_i$ is the traction vector and W is the strain-energy density. The *J* line integral is path-independent for nonlinear elastic materials deforming according to the deformation theory of plasticity, and *substantially* path-independent for incrementally deformed plastic materials conforming to the flow theory of plasticity. It can be shown that *J* is identical with the strain-energy release rate for linear elastic fixed-grip conditions. The *J*-integral derived from *nonlinear* elasticity theory is a similar function to the strain-energy release rate of linear elastic fracture mechanics. The *J*-integral is confined to materials that exhibit reversible nonlinear elastic deformation. Because plastic deformation is irreversible, it is found that aspects of the derivation of the integral in the fully plastic case are invalid. For example, it is not possible to prove that $J = 0$ for a closed loop, as described by Ewalds and

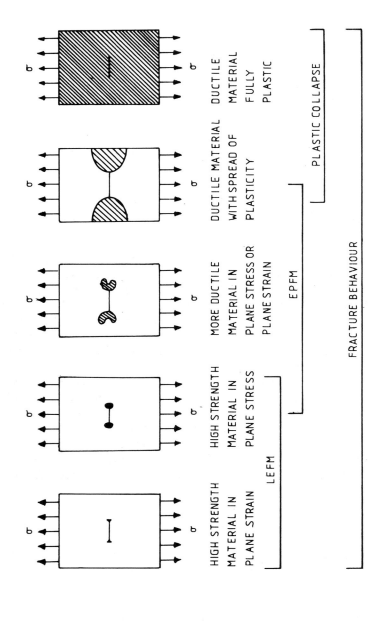

Fig. 1.7 Ranges of application of LEFM and EPFM. (After H. L. Ewalds and R. J. H. Wanhill, *Fracture Mechanics*. Arnold/DUM, London/Delft, 1984.

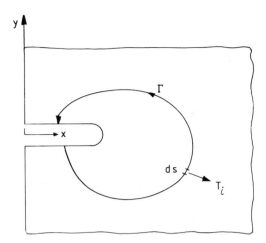

Fig. 1.8 Closed circuit for the calculation of the J-integral.

Wanhill (1984). Also, it can be argued that J can only be used up to the beginning of crack extension; once the crack begins to extend, unloading of the material occurs in that region. Nevertheless, elasto-plastic materials can be modelled using the J-integral concept with its assumption of nonlinear elasticity, as long as no unloading occurs. Because of these uncertainties in the application of the J-integral to the large plastic deformations that occur in ductile fracture, there have been few attempts to utilize this approach under these circumstances.

Another approach to the onset of crack extension was made by Wells (1962), who introduced the crack opening displacement. Here a critical crack opening displacement corresponds to the condition for crack extension, which in turn is a characteristic of the crack-tip stress field. Burdekin and Stone (1966), using the well-known model derived by Dugdale (1960), which assumes a crack with a tip plastic field, which may be replaced by a crack with an effective length that includes no plasticity, found that the crack opening displacement δ was given by

$$\delta = \frac{8\sigma_y a}{\pi E} \ln \sec \left(\frac{\pi \sigma}{2\sigma_y} \right). \qquad (1.8)$$

Here a is the real crack length, σ_y is the yield strength and σ is the applied stress. Now, if we consider $\sigma \to \sigma_y$ then $\sec (\pi\sigma/2\sigma_y) \to \infty$, and therefore it follows that $\delta \to \infty$. So, under conditions of general yielding the use of COD becomes difficult.

Crack extension or propagation is assumed to occur when the COD reaches a critical value. In the case of linear elastic fracture mechanics, this is equivalent to a critical K or a critical strain-energy release-rate criterion. But it is unclear what approximations must be made to use COD under conditions of large plastic strains that pertain in ductile fracture.

1.8   Microstructural features

The structure of engineering alloys can be highly complex as indicated in Fig. 1.9. Coherent and incoherent precipitates are found in age-hardenable materials such as aluminium alloys, and the rod-shaped second-phase particle is a characteristic feature of some intermetallic compounds. Grain-boundary precipitates are potentially sites of weakness, for example the grain-boundary precipitates of cementite in hypereutectoid steel. The two types of fracture are classified in Table 1.2; these are transgranular and intergranular. Examples of transgranular fractures are void coalescence, cleavage fracture and fatigue crack initiation and growth. Intergranular fractures include grain-boundary separation with and without void coalescence. Further types of fracture are creep fractures, which occur at elevated temperature, and stress corrosion cracking. Ductile transgranular fractures are caused by void nucleation at second-phase particles or grain boundaries, and void coalescence occurs with increasing applied strain.

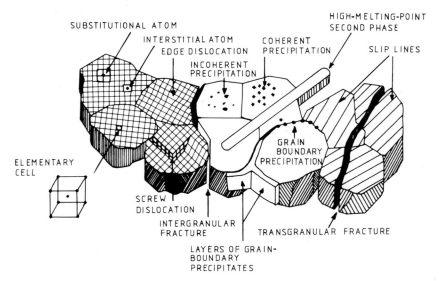

Fig. 1.9 Microstructural features in metallic materials (After H. L. Ewalds and R. J. H. Wanhill, *Fracture Mechanics*. Arnold/DUM London/Delft, 1984; also after Gerling Institüt für Schadensforschung.)

Table 1.2

Important kinds of transgranular fracture

Fracture	Typical occurrence	Characteristics
Microvoid coalescence (ductile fracture)	Stable and unstable fracture High-stress-intensity fatigue	Irregular blunt crack fronts Considerable local plasticity
Cleavage in steels	Unstable fracture Intermediate- and high-stress-intensity fatigue Hydrogen embrittlement	Sharp kinked cracks Brittle fracture
Continuum fatigue (striations)	Intermediate- and high-stress-intensity fatigue	Uniform crack fronts; Limited blunting Mode I favoured Considerable local (cyclic) plasticity
Slip-plane cracking (faceted fracture)	Fatigue crack initiation and growth at low stress intensities, especially under torsional loading	Branched, kinked and irregular crack fronts, limited blunting Mode II favoured Limited local (cyclic) plasticity
Cleavage-like faceted fracture	Low-stress-intensity fatigue Stress corrosion	Branched, kinked and irregular crack fronts, limited blunting Limited local plasticity

Classification of intergranular fracture

Fracture	Typical occurrence	Mechanical characteristics
Microvoid coalescence along grain boundaries	Stable and unstable fracture and high-stress-intensity fatigue in high-strength steels and aluminium alloys	Limited local plasticity
Low-ductility grain-boundary separation	Sustained load fracture (including creep)	Little evidence of plasticity
Brittle grain-boundary separation	Unstable fracture in temper-embrittled steels and refractory metals (e.g. tungsten)	Brittle low-energy fracture

Adapted from H. L. Ewalds and R. J. H. Wanhill *Fracture Mechanics*. Arnold/DUM, London/Delft, 1984.

1.9   Crack advance

As we have mentioned previously, there are three basic features of fracture in simple tension; these are cleavage, void formation and rupture. These features are also found in more complex stress states. Thus any ductile fracture occurring in engineering practice can be divided into several basic features signified by these three distinct modes. In the case of crack advance, Ashby (1981) classified modes of crack advance in terms of the above three features. By using the fracture toughness, which is strictly only suitable for elastic or elasto-plastic regimes and brittle fracture, Ashby arrived at

$$K_{c} \sim \tfrac{1}{10}E(2\pi r^{*})^{1/2} \tag{1.9}$$

for brittle fracture without plastic deformation at the crack tip and

$$K_{c} \sim (2\pi\sigma_{y}r^{*}E\varepsilon_{f})^{1/2} \tag{1.10}$$

for ductile fracture with crack-tip plasticity, where K_{c} is the fracture toughness, E and σ_{y} are Young's modulus and the yield stress respectively, ε_{f} is a critical strain and r^{*} is a characteristic distance. A similar expression has been given by Cheng (1982).

Owing to limited plasticity for cleavages 1 and 2, equation (1.9) should be suitable, and the corresponding characteristic distance r^{*} should be the atomic spacing and the grain size respectively. If the plasticity is fully developed ahead of the crack tip then the material fractures in a ductile manner and the characteristic distance should be the inclusion spacing or void spacing, provided that the fracture is void-controlled.

In the case of pure metals in which rupture occurs r^{*} is the sheet width in plane-stress testing, which is certainly significantly larger than characteristic microstructural distances. Therefore this behaviour leads to a high toughness, as shown in Table 1.3, which gives typical values of toughness from $0.2 \text{ MPa m}^{1/2}$ for brittle fracture to $2000 \text{ MPa m}^{1/2}$ for rupture-controlled ductile fracture in pure metals. These are in agreement with the usually accepted values. This demonstrates that toughness is controlled by the three previously mentioned modes of fracture, and these modes do represent the basic features of ductile fracture.

1.10   Combination of fracture modes and fracture maps

In practice the three modes of ductile fracture are usually combined. For example, Fig. 1.10 shows a copper tensile specimen that clearly shows these

Table 1.3

Magnitude of G_c and K_c and their relationship to the mechanisms of crack advance (after M. F. Ashby, *Prog. Mat. Sci.; Chalmers Anniversary Volume*, pp. 1–25 (1981))

	Toughness G_c (J m^{-2})	Fracture toughness K_c (MPa m$^{1/2}$)	G_c/Eb	$K_c/Eb^{1/2}$
Plastic rupture	10^6–10^7	500–2000	$\sim 10^5$	>100
Fibrous by void growth and coalescence	10^4–10^6	20–500	10^2–10^4	10–100
Unstable shear void-sheet linkage	10^3–10^5	10–100	$\sim 10^3$	2–20
Mixed fibrous and cleavage	10^3–2×10^4	10–50	$\sim 10^2$	2–10
Cleavage preceded by slip (cleavage or BIF 3)	30–300	2–20	1–10	0.5–5
Cleavage without slip (cleavage or BIF 1 and 2)	1–100	0.2–5	0.1–1	0.1–2
2 × (surface energy), metals	0.5–3	0.4–1	0.1	~0.2
2 × (surface energy), ceramics	0.5–4	0.2–1	0.1	~0.2
2 × (surface energy), polymers	0.1–1	0.01–0.05	0.1–10	~0.2

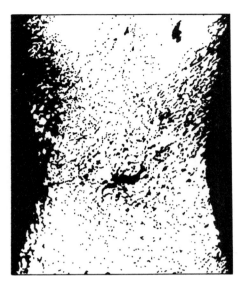

Fig. 1.10 Section of a necked copper specimen showing voids. (After K. E. Puttick, *Phil. Mag.* **4**, 964–969 (1959).)

combinations. This is a very famous photograph by Puttick (1959), being one of the first to illustrate void growth in a neck. However, as reported by Nadai (1950), MacGregor (1937) verified that the crack begins in the centre of the neck by using X-ray photographic techniques. In the central part of the neck there is a crack surrounded by several cavities, which have most probably nucleated at inclusions. Some voids are large, revealing typical void growth in tension. Between the voids and ahead of the central crack there are regions of intensive shear, which will, in turn, promote void coalescence and extend the central crack. The zones of intense shear appear to be the cause of the zig-zag path of the crack. One advantage of this particular example is that it indicates the true complexity of ductile fracture.

Depending on the environment and loading conditions, fracture can change from one mode to another. There are a number of approaches for aiding the description of these fracture transitions. The fracture maps described by Frost and Ashby (1982) illustrate the transitions well. Figure 1.11 is the fracture-mechanism map for magnesium oxide in tensile tests. This figure shows virtually all of the fracture modes mentioned above. The axes of these fracture-mechanism maps are normalized tensile stress and homologous temperature. The region labelled dynamic fracture implies fractures accompanying stress wave propagation.

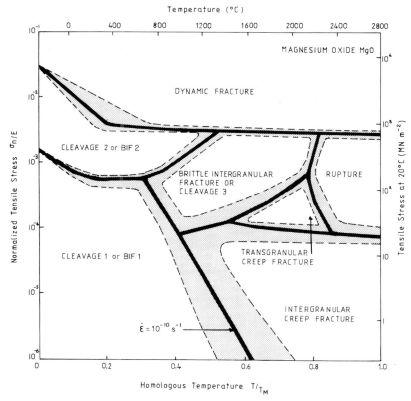

Fig. 1.11 Fracture mechanism map for MgO. (After M. F. Ashby, *Prog. Mat. Sci.; Chalmers Anniversary Volume*, pp. 1–25 (1981).)

1.11 Mechanisms related to ductile fracture

Although the three modes of ductile fracture represent fundamental processes and possess different features, on examining them in detail from the physical as opposed to the phenomenological point of view we can identify three separate mechanical features that underlie ductile fractures. There are:

(i) plastic flow concerning stress concentration and localization of deformation;

(ii) void dynamics, namely nucleation, growth and coalescence;

(iii) localization of plastic deformation, particularly the two main types of localization: necking and shearing.

The first of these features is concerned with constitutive relationships in the plastic regime. Some plastic-flow results will be used frequently in this book. For further details of constitutive equations it is recommended that readers consult the textbook by Malvern (1969) and the monograph by Campbell (1973). As we have seen, features (ii) and (iii) may be regarded as the physical mechanics of ductile fracture in metals. Therefore these two topics will inevitably form the core of the book.

1.12 Metalworking

The final part of the book is devoted to metalworking processes and limitations on the processes resulting from ductile fracture. It is felt that this important industrial area will place the theories of ductile fracture in an appropriate context. The application of theories of flow localization and void dynamics to workability limits in sheet and bulk working processes and to the mechanics of metal cutting are described. It should be borne in mind that both the theories of ductile fracture and/or the understanding of these metalworking processes are still progressing. Hence the correlation of theories with practical processes has necessarily to be elementary, even inconsistent in some cases. Despite this, the governing mechanisms are becoming clearer. This will be of benefit to manufacturing industry, with its large volume production.

Finally, unless otherwise stated, elastic strains are assumed to be negligibly small throughout this book.

2

Measures of Ductility

A metal that plastically deforms extensively without the onset of fracture is normally termed ductile. One of the intriguing problems with the ductility of metals is that ductility can be very sensitive to, among other things, the applied stress state. For example the axial fracture strain in a tensile test will differ significantly from the shear strain at fracture obtained in a torsion test. To illustrate these differences three types of test are described here: tension, compression and torsion tests. The different measures of ductility derived from each test are described, as well as the underlying complexity of each experimental method.

2.1 Tensile tests

In a tensile test the specimen is subjected to a tensile force along its length. Specimens, in the form of cylindrical rods or flat sheets, usually have a reduced section, or gauge length, the purpose of which is to ensure a uniform stress within the specimen. Test specimens have standardized dimensions.

The smooth curve shown in Fig. 2.1 is typical of the stress–strain behaviour of most metals and alloys. The engineering stress is defined as the load divided by the initial cross-sectional area of the specimen, and the engineering strain is defined as the change in length divided by the original gauge length.

Usually, the yield stress is difficult to find accurately from such a curve; therefore an offset yield strength is often quoted. The ultimate tensile strength corresponds to the maximum in the applied load experienced by the material. The uniform strain is simply the strain corresponding to the ultimate tensile strength. At the peak load a neck develops, and with subsequent straining the load decreases until fracture occurs.

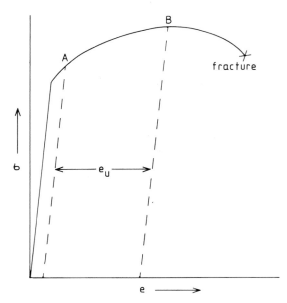

Fig. 2.1 Typical engineering stress σ versus engineering strain e for a nonferrous metal. A is the offset yield stress, B the ultimate tensile stress and e_u is the uniform strain.

A number of materials, typically numerous steels, exhibit discontinuous yielding, as shown in Fig. 2.2. The highest stress A before the sudden yield drop is called the upper yield stress, and the approximately constant stress after the drop is called the lower yield stress B (see Fig. 2.2). Both the upper and lower yield stresses increase with decreasing temperature and increasing strain rate. After the initial drop in stress, a narrow band of plastically deformed material is nucleated, which forms at a characteristic angle to the tensile axis. This band is called a Lüders band. The band grows in width until the stress required for the nucleation of a second band is reached. This type of behaviour is repeated until the entire gauge length becomes plastic. Each "blip" at the lower-yield-point level corresponds to the nucleation and growth of individual bands. Nadai's (1950) book contains some excellent examples of Lüders bands formed in steel sheets, and is recommended for further reading on the subject.

Two types of dislocation theory have been proposed to explain discontinuous yielding. Cottrell and Bilby (1949) suggested that discontinuous yielding is caused by interstitial carbon (and nitrogen) atoms. It is energetically favourable for the interstitial atoms to migrate to the dislocations, anchoring, or pinning, them. The upper yield stress is considered to

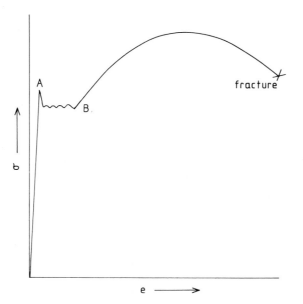

Fig. 2.2 Engineering stress–strain curve for a mild steel. *A* is the upper yield stress and *B* is the lower yield stress.

be the stress at which dislocations may be freed from their interstitial atmospheres, and the lower yield stress is the stress required to make the dislocations glide along their slip planes.

An alternative explanation was suggested by Johnston and Gilman (1959). They proposed that the sudden decrease in stress is caused by the rapid multiplication of a large number of fresh dislocations from Frank–Read and other suitable sources.

When the entire gauge length is plastic, that is for strains greater than the Lüders strain, the specimen strain hardens uniformly, and the engineering stress–strain curve is similar to that shown in Fig. 2.1.

Either the total elongation to fracture or the reduction in area at fracture measured in the neck are normally taken as measures of ductility in tensile tests. The two quantities are defined as

$$e = \frac{l_{\mathrm{f}} - l_0}{l_0}, \quad R = \frac{A_0 - A_{\mathrm{f}}}{A_0},$$

where l_0 and l_{f} are the initial and final gauge lengths respectively, and A_0 and A_{f} are the initial and final cross-sectional areas. These measures of ductility are discussed in greater detail in §2.1.5.

2.1.1   Maximum load in tension

For ductile metals subjected to plastic deformation in tension the load-carrying capacity of the test-piece first rises, reaches a maximum and then decreases with increasing strain. At the load maximum the increase in the flow stress caused by strain hardening is just offset by the decrease in load-carrying area, this decrease is often referred to as geometrical softening. At this maximum in the load, necking is initiated.

The condition for a load maximum is given simply by

$$dL = 0. \tag{2.1}$$

Now the load

$$L = \sigma A, \tag{2.2}$$

where A is the current area and σ is the true stress and therefore $dL = \sigma dA + A\,d\sigma$. Because the volume is assumed to remain constant during the plastic deformation, $d\varepsilon = -dA/A$, the load maximum corresponds to the condition

$$\frac{d\sigma}{d\varepsilon} = \sigma. \tag{2.3}$$

A schematic representation of this condition is shown in Fig. 2.3.

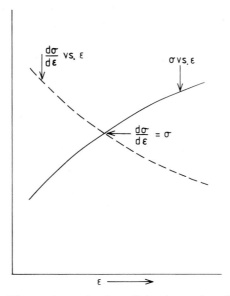

Fig. 2.3   The maximum load condition in tension, $d\sigma/d\varepsilon = \sigma$.

A much earlier geometrical construction that provides the maximum load point is due to Considère (1885). This involves plotting the true stress against the engineering strain; see Dieter (1976) for a description of the construction.

If the true stress–strain curve of the material can be described by the Ludwik-type power-law strain-hardening equation

$$\sigma = K\varepsilon^n \tag{2.4}$$

then it may be shown simply that the maximum load condition occurs when the uniform strain ε_u is given by the strain hardening exponent

$$\varepsilon_u = n. \tag{2.5}$$

2.1.2 Onset of localized necking in sheet tensile specimens

When an annealed sheet-metal tensile specimen is tested it becomes unstable in the same way as for rod specimens. This is called diffuse necking. As the strain is increased, the neck gradually develops into what eventually becomes a localized neck. In localized necking the specimen thins along a band inclined at an angle ϕ to the tensile axis, as shown in Fig. 2.4. Plane-strain conditions prevail in the neck and in order for the neck to develop the normal strain along the length of the band must be zero. Hence $\phi = \frac{1}{2}\{\pi - \cos^{-1}(\frac{1}{3})\} = 54.7°$ (Fig. 2.4), since $\varepsilon_2 = \varepsilon_3 = -\frac{1}{2}\varepsilon_1$.

The rate of strain hardening at the onset of localized necking is cancelled by geometrical softening within the band or neck, i.e. thinning. Since the area

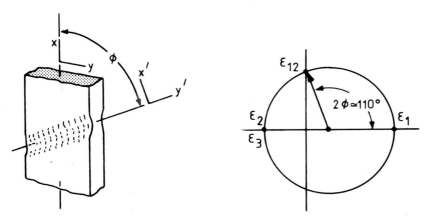

Fig. 2.4 Localized necking in thin-sheet tensile specimens. The angle ϕ that the neck makes with the tensile axis is determined by the requirement for zero strain increment along the length of the band. For an isotropic material $\phi = \tan^{-1}\sqrt{2} \approx 54° \ 44'$.

is only changed in one dimension,

$$dA = \omega dt, \tag{2.6}$$

where dA is the increment in area, w is the width of the band and dt is the thickness increment. It follows that

$$\frac{d\sigma}{d\varepsilon} = -\frac{\sigma \, dt/t}{d\varepsilon}, \tag{2.7}$$

and until the neck appears

$$\frac{dt}{t} = d\varepsilon_3 = -\frac{d\varepsilon}{2}. \tag{2.8}$$

Substituting (2.8) into (2.7) gives as the condition for the onset of localized necking

$$\frac{d\sigma}{d\varepsilon} = \frac{\sigma}{2}. \tag{2.9}$$

In terms of the strain-hardening exponent n, assuming again for simplicity that the stress–strain curve can be represented by the power law $\sigma = K\varepsilon^n$, the onset of localized necking occurs when

$$\varepsilon = 2n. \tag{2.10}$$

Localized necking may also be considered more formally as a problem of plane-stress characteristic theory (see Hill, 1952). It may be shown that the angle between the length of neck and the tensile axis is a function of the plastic anisotropy of the sheet, as described by Hill (1950). This formal approach to localized necking will be returned to in Chapter 7, in which localized shearing in plane strain is considered.

2.1.3   Stress state in the neck

In ductile metals after the maximum in the applied load a neck forms at some position along the gauge length. Further plastic deformation is concentrated in the necked or constricted region. In some polymers localization of deformation to a necked region is metastable and considerable plastic deformation can occur as the neck gradually spreads along the gauge length, as discussed by Nadai (1950) in relation to nylon. However, in most metals fracture eventually occurs in the necked region, the neck not being stable.

Once a neck occurs in a rod specimen, a rotationally symmetrical triaxial tensile stress state is induced in the region, this is shown schematically in Fig.

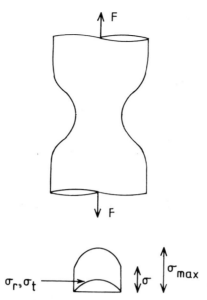

Fig. 2.5 Rotationally symmetric stress field in the region of a neck. σ_{max} is the peak axial stress, σ is the true stress in uniaxial tension and σ_r and σ_t are the radial and tangential stresses respectively.

2.5. The tensile stresses have their highest values at the midpoint of the minimum section. With continued straining, the neck becomes more pronounced and the radial and transverse stresses increase in magnitude. The average axial stress, defined here as the axial load divided by the minimum cross-section of the neck, is higher than the stress with no neck. This indicates that a neck behaves in a similar manner to a notch—it provides a plastic constraint. Plastic constraint arises from the triaxial stress state at the notch root. When triaxial stresses are present, the yield stress for the notched section will be higher than the uniaxial yield stress by some factor. This factor is commonly referred to as the plastic constraint factor. The fact that the hardness of a metal is between two and three times its uniaxial yield stress is due to the plastic constraint around the indentation caused by the surrounding elastic material. The constraint factor in the case of a neck can be defined as the average axial stress divided by the axial stress acting on a uniform rod of the same diameter as the plane of minimum section.

It is possible to estimate the average axial stress, compensating for the induced radial and transverse stresses. If it is assumed that the neck profile has a characteristic radius R and the radius of the minimum section is a then

Bridgman (1944) has provided a correction factor. Bridgman's analysis is based upon the following assumptions:

(i) the contour of the neck has a radius that is a characteristic of the material and the testpiece geometry;
(ii) the plane of minimum section of the neck remains circular during plastic deformation;
(iii) the von Mises yield criterion is applicable;
(iv) the strains are constant over the cross-section of the neck.

The above assumptions lead to the following expression for the corrected axial stress:

$$\sigma = \frac{\sigma_{av}}{(1 + 2R/a) \ln \{1 + a/(2R)\}}, \qquad (2.11)$$

where σ_{av} is the axial load divided by the minimum cross-sectional area. The radii a and R can be obtained either by straining a specimen to given elongations and measuring a and R, or alternatively they can be measured from photographs taken during the test.

There have been other attempts to take into account the nonuniform triaxial tension in the neck, for example the work of Davidenkov and Spiridonova (1946). However, it is now generally accepted that Bridgman's analysis is the more accurate.

Tegart (1966) plotted the stress ratio σ/σ_{av} against the strain in the neck minus the strain at necking for copper, steel and aluminium; these results are shown in Fig. 2.6. The theoretical plot according to Bridgman is shown by the solid line. A good correlation is obtained. Tegart subtracted the strains at necking to take account of the widely different necking strains for the three materials considered.

2.1.4   Crack initiation sites in rod specimens

Because of the large triaxial tensile stress state induced at the minimum section of the neck, it is not surprising that cracks have been observed to initiate in the same region. It appears that Ludwik (1927) was the first to observe that fracture in rods begins on the minimum section of the neck. MacGregor (1937) verified that fracture in rods of aluminium begins with a crack at the centre of the neck. MacGregor used X-ray photographs at various stages in the process, and from these he was able to observe the cracks soon after initiation. These cracks have been observed by MacGregor and others to zig-zag to the surface. This and other aspects of neck geometry are described in Nadai (1950).

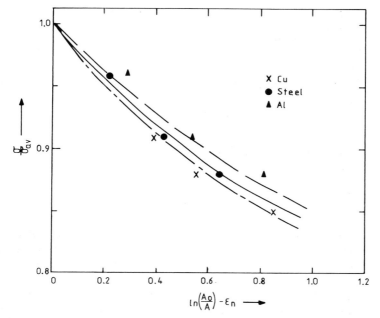

Fig. 2.6 Correction factor for necking as a function of the strain in the neck minus the strain at necking. (Reprinted with permission of Macmillan Publishing Co. from *Elements of Mechanical Metallurgy*, by W. J. M. Tegart. Copyright 1966 by W. J. McGregor Tegart.)

2.1.5 Measures of ductility

After initiation of necking, plastic deformation is confined to the necked region and the applied load continues to drop until the specimen fractures. The final extension consists of two components: the uniform extension up to the necking point and a localized extension that is concentrated in the neck. A typical variation of local elongation with position along the gauge length is shown in Fig. 2.7.

The extension of the specimen at fracture is normally written as (Tegart, 1966).

$$l_f - l_0 = \alpha + \beta l_0, \qquad (2.12)$$

where α is the local extension, βl_0 is the uniform extension and l_0 and l_f are the initial and final gauge lengths respectively. From the definition of engineering strain,

$$e_f = \frac{l_f - l_0}{l_0} = \frac{\alpha}{l_0} + \beta. \qquad (2.13)$$

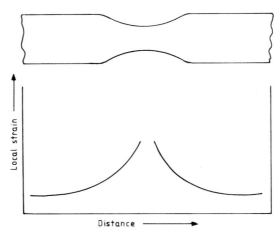

Fig. 2.7   Variation of strain along the gauge length of a specimen.

It follows that the engineering fracture strain is not an absolute material property for a given strain rate and temperature, but can vary with the gauge length.

The uniform extension is independent of the specimen area, but the local extension in the neck is affected by the cross-sectional area.

It is generally assumed that geometrically similar specimens develop geometrically similar necks. If we use Barba's (1880) law $\alpha = cA_0^{1/2}$ then the so-called elongation equation becomes

$$e_f = \frac{cA_0^{1/2}}{l_0} + \beta. \tag{2.14}$$

If this equation is accurate then for geometrically similar bars constant $cA_0^{1/2}/l_0$ must be used. Nevertheless, it can be seen that the fracture strain depends on $A_0^{1/2}/l_0$.

A strikingly simple indication of the geometry dependence has been discussed by Rogers (1968). Consider a rod tensile specimen that fractures due to rupture (Fig. 2.8a). If a bundle of untwisted strands of wire that has the same total cross-sectional area as the rod is now made up and this specimen is pulled to fracture (Fig. 2.8b) then it will be observed that each strand will have 100% reduction in area but the apparent reduction in area of the bundle will be markedly smaller.

Thus a conclusion that can be drawn from this section is that neither the elongation to fracture nor the reduction in area at fracture are fundamental

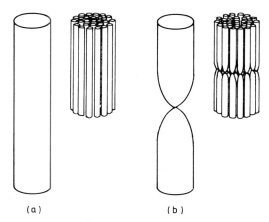

(a)   (b)

Fig. 2.8   The effect of specimen geometry on ductility. (After H. C. Rogers, in *Proc. ASM Ductility Seminar*, pp. 31–61. ASM, Ohio, 1968.)

measures of ductility. However, for uniaxial tensile tests the reduction in area is more significant than the extension to fracture.

2.1.6   Effect of superimposed hydrostatic pressure on fracture in rods

Bridgman (1952) showed that the tensile fracture of steel-rod specimens could change from cup-and-cone fractures to rupture if the superimposed hydrostatic pressure was high enough. Further details of the effects of superimposed hydrostatic stress on ductility are given in Chapter 3.

2.2   Compression

In general no difficulties are encountered in subjecting a rod or a sheet specimen of a ductile material to a uniform tensile force, at least until the onset of necking. However, it is more difficult to ensure homogeneous deformation in compression testing because of the presence of friction at the die–test-piece interfaces. As the material is compressed, fresh material will come into contact with the dies as compression continues. Clearly, the dies need to be well lubricated. As the compressive axial strain is increased, the test-piece increases in cross-sectional area. Therefore there is no possibility of geometrical instability (necking), as there is in tensile tests. Compression tests are particularly valuable for obtaining the stress–strain behaviour of materials at high strains because of the suppression of the geometrical instability.

2.2.1   Uniaxial compression of cylinders

Conventional specimens for compression are cylinders with aspect ratios (height–diameter ratios) less than about 2, so that sideways buckling does not occur. To ensure that the cylinder expands uniformly, the test is usually carried out incrementally. After each load increment, the test-piece is unloaded, fresh lubricant is applied and then is reloaded. The load–height-reduction relation and therefore the stress–strain curve can be derived from measurements of the change in height and load.

If the test-piece is deformed homogeneously then the true stress is given simply by

$$\sigma = L/A_i, \tag{2.15}$$

where A_i is the current cross-sectional area. The true height strain is

$$\varepsilon = \ln (h_0/h_i) \tag{2.16}$$

(by convention is compression testing, compressive strains are taken to be

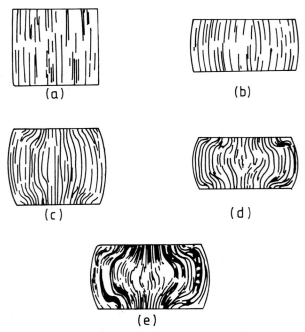

(a)   (b)

(c)   (d)

(e)

Fig. 2.9   Effect of friction on deformation of cylinders in compression. (Based on a drawing by A. Nadai, *Theory of Flow and Fracture of Solids.* McGraw-Hill, New York, 1950.)

Fig. 2.10 Base of a cylinder after compression, showing distortion of streaks of impurities. The outer ring of material originally constituted the cylindrical surface of the specimen. (Based on a drawing by A. Nadai, *Theory of Flow and Fracture of Solids.* McGraw-Hill, New York, 1950.)

positive). Lastly, another commonly quoted quantity is the reduction in height

$$r = \frac{h_0 - h_i}{h_0}.\qquad(2.17)$$

Figures 2.9(a)–(e) are sections of steel cylinders showing the changes of initially parallel lines of impurities (the original sketches were made by Nadai (1950)). The less severe the frictional constraint at the ends, in other words the better the lubrication, the less severe is the barrelling. Cylinders lubricated well with graphite expand uniformly or nearly so (Figs. 2.9a–c), whereas cylinders compressed without lubricant exhibit severe barrelling (Figs. 2.9d, e). Conical dead metal caps form under the dies when no lubricant is used. The shear zones producing these conical surfaces are continually reforming as the test-piece is compressed.

It was observed by Nadai (1950) that the ends of the cylinder tend to invert and move so that the material becomes part of the outermost ring, as shown in Fig. 2.10.

2.2.2 Plane-strain compression

The object of this test is to obtain the stress–strain behaviour of the test material with one of the strains, ε_2 in Fig. 2.11, constrained to be zero.

The dies overlap the strip and are narrow, ensuring that the plastic region remains small and is directly between the dies. With incremental loading and repeated lubrication accurate stress–strain curves are obtainable using this technique.

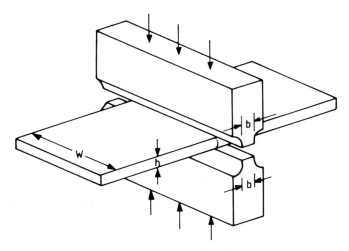

Fig. 2.11   Plane-strain compression dies and specimen.

Watts and Ford (1952) have carried out extensive tests on copper sheet and have perfected the method. They recommend the following:

(i) incremental loading in the same identation,
(ii) the ratio of die breadth to strip thickness should be between 2 and 4,
(iii) the width of the strip should be at least six times the die breadth.

2.2.3   Modes of fracture in compression

If deformation is frictionless then fracture is not expected in compression of cylinders. However, catastrophic modes of fracture in which the specimens break into two or more parts have been observed by Latham (1963) and Dodd *et al.* (1985). Latham's tests were carried out on brass and fracture occurred after reductions in height of about 30%. On the other hand the experiments of Dodd *et al.* were carried out on a magnesium alloy, and fracture occurred at reductions of about 15%.

Photographs of fractured specimens tested by the latter workers are shown in Fig. 2.12. For specimens with aspect ratios of unity or below, fracture occurred by the propagation of cracks from one end to the other. For specimens with aspect ratios greater than unity, fracture occurred between the ends and approximately the midheights. Interestingly, no barrelling was observed in these magnesium tests, no matter what conditions of lubrication were used.

In more ductile materials two modes of cracking have been observed by

Kudo and Aoi (1967) beginning at the equator. For specimens subjected to severe end constraint (and therefore severe barrelling) vertical cracks occur, whereas for well-lubricated tests with little barrelling oblique cracks were observed at the equator. These two modes of cracking are discussed in greater detail in Chapter 8.

2.2.4 Thermal localization in forging

In a series of elevated temperature forging experiments, Johnson *et al.* (1964) observed what they called lines of thermal discontinuity. These lines are of importance for two reasons. First, they demonstrate that intense shear and inhomogeneous deformation occur in practical forging operations. Secondly, the lines show that high local temperatures can develop along the intense shear bands.

The lines of thermal discontinuity, also called thermal crosses by Johnson *et al.*, follow the patterns to be expected from upper-bound analyses. There are excellent coloured photographs of thermal discontinuities in the paper by Johnson *et al.* (1964).

It is interesting to note that such lines were reported as early as 1921 by Massey from forging experiments carried out on steels at about 680 °C (see Slater, 1965/6). Local temperature rises of between 50 and 100 °C have been calculated in the region of these lines of intense shear.

2.3 Torsion

Torsion tests on thin-walled circular tubes are now accepted as an accurate method for the determination of stress–strain characteristics of materials. As there is no change in cross-sectional area during twisting, necking does not occur and it is possible to test to larger strains than in tension. These strains are often comparable to, or approaching those, observed in metal-forming processes. Further, torsion tests may now be carried out at relatively high strain rates using stress-wave loading techniques.

2.3.1 Twisting of rods and tubes

Twisting of plates and cylindrical rods are extremely complex tests to analyse, as has been described by Nadai (1950).

For twisting of a solid rod the shear strain is given by

$$\gamma = r\theta, \tag{2.19}$$

where r is the radial distance from the specimen axis and θ is the angle of twist

(a)

(b)

per unit gauge length of the specimen. By considering the rod to be made up of a series of thin-walled tubes, the total torque is found to be

$$M = 2\pi \int_0^{r_s} \tau r^2 \, \mathrm{d}r, \qquad (2.20)$$

where r_s is the radius of the rod and τ is the shear stress corresponding to the

(c)

Fig. 2.12 General views of fractures of magnesium alloy compression specimens: (a) aspect ratio $\frac{2}{3}$; (b) aspect ratio 1; (c) aspect ratio $1\frac{1}{2}$. The material fractured in an intergranular manner. For aspect ratios less than unity, fractures propagate from one end of the specimen to the other; whereas for aspect ratios greater than unity, fractures are confined to one end. (After B. Dodd *et al.*, *Res. Mechanica*, **13**, 265–273 (1985).)

radius. By substitution, this equation becomes

$$M = \frac{2\pi}{\theta^3} \int_0^{\gamma_s} \tau \gamma^2 \, d\gamma. \tag{2.21}$$

If we assume that the shear stress in the specimen is only dependent on the shear strain then the following equation can be derived for the surface shear stress, after Nadai (1950), by differentiating (2.21) and taking $\gamma_s = r_s\theta$:

$$\tau = \frac{1}{2\pi r_s^3}\left(3M + \theta \frac{dM}{d\theta}\right). \tag{2.22}$$

Fields and Backofen (1957) derived the following equation for the shear stress of rate-sensitive materials:

$$\tau = \frac{M}{2\pi r_s^3}(3 + n + m), \tag{2.23}$$

where n and m are the slopes of the log M–log θ plots at constant $\dot{\theta}$ and log M–log $\dot{\theta}$ plots at constant θ.

A further pair of formulae can be derived if the so-called differential test is used; here two specimens with slightly different radii are used. The formulae for this case are

$$\tau_d = \frac{3(M_2 - M_1)}{2\pi(r_2^3 - r_1^3)}, \quad \gamma_d = \left(\frac{r_1 + r_2}{2}\right)\theta, \tag{2.24}$$

where r_2 and r_1 are the radii and $r_2 > r_1$.

Unlike torsion tests on solid rods, torsion tests on thin-walled tubes are not so difficult to analyse because there is no restriction on the strain-hardening characteristics of the material.

The torque in a specimen with inner radius r_i and outer radius r_o is given by

$$M = 2\pi \int_{r_i}^{r_o} \tau(r)r^2 \, dr. \tag{2.25}$$

Because we can assume that the shear stress does not vary across the wall the average shear stress $\bar{\tau}$ is

$$\bar{\tau} = \frac{3M}{2\pi(r_o^3 - r_i^3)}, \tag{2.26}$$

or, in terms of the mean radius r_m,

$$\tau_m = \frac{M}{2\pi r_m^2 t}, \tag{2.27}$$

where t is the specimen thickness.

Recently a number of error-estimation techniques for torsion have been discussed, and one technique has been proposed by Kobayashi and Dodd (1985) for estimation of errors in thin-walled tube tests.

In the case of elastic deformation, $\tau = Gr\theta$, and from (2.25) the actual stress at the mean radius is

$$\tau_a = \frac{M}{\pi(r_o - r_i)(r_o^2 - r_i^2)}, \tag{2.28}$$

where r_o and r_i are the outer and inner radii respectively. The ratio between this stress and τ_m given by (2.27) is

$$\frac{\tau_a}{\tau_m} = \frac{(a + 1)^2}{2(a^2 + 1)}, \tag{2.29}$$

where $a = r_o/r_i$. For example, for a wall thickness of 0.5 mm and an inner radius of 5 mm, $\tau_a/\tau_m = 0.9977$. Thus, even in the case of a wholly elastic specimen, (2.27) provides an accurate measure of the actual shear stress at the mean radius.

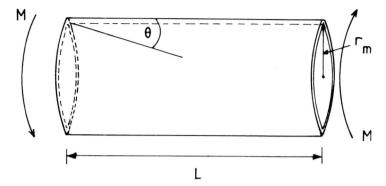

Fig. 2.13 Twisting of a thin-walled tube.

Specimen dimensions for torsion testing of thin-walled tubes have been discussed by Clyens (1973).

Referring to Fig. 2.13, the shear strain in the tube is given by

$$\gamma = r_m \theta. \tag{2.30}$$

2.3.2 Shear localization

In a rod tensile test the load-carrying capacity of the specimen reaches a maximum with increasing axial plastic strain owing to the decrease in cross-sectional area. At this maximum a further increment in the applied load causes a neck to form, as we have already seen in §2.1.1. In a torsion test on a thin-walled tube, on the other hand, there is no change in area of the specimen, and therefore the geometrical effect of necking is absent.

Even in torsion, however, a maximum in applied torque is eventually reached. As there is no possibility of geometrical softening, the maximum in the torque is attributed to thermal softening, which further localizes flow (see e.g. Lindholm *et al.*, 1980; Dodd and Atkins, 1983a). As the plastic strain increases, the temperature of the specimen increases, the amount of heat produced depending on the specimen dimensions, the applied strain rate and the mechanical and physical properties of the test material. For constant specimen dimensions, as the strain rate increases, the strain at which shear localization occurs will decrease.

Figure 2.14 shows some torsion-test results on copper carried out at strain rates between 0.009 and 330 s^{-1}. For strain rates between 0.009 and 9.6 s^{-1} no stress maxima are observed and no fracture occurred. However, for strain rates between 174 and 330 s^{-1} shear localization occurs. Further, in these

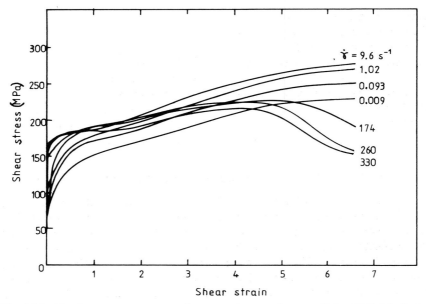

Fig. 2.14 Torsion tests on copper tubes at various strain rates. (After U. S. Lindholm *et al.*, *Trans. ASME*, H: *J. Engng Mat. Technol.* **102**, 376–381 (1980).)

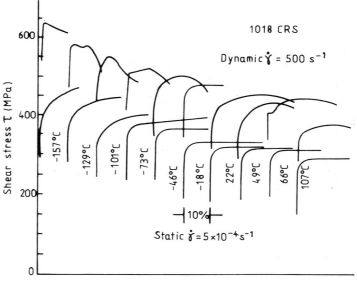

Fig. 2.15 Tests carried out at various temperatures and two strain rates on 1018 steel. (After Costin *et al.*, *Inst. Phys. Conf. Ser.* No. 47 (ed. J. Harding), pp. 90–100 (1979).)

higher-strain-rate tests the rates of strain hardening are lower than those for the lower-strain-rate tests.

In the test carried out at 1 s^{-1} it was found that the gauge length had been subjected to uniform plastic deformation. For the higher-strain-rate tests central bands of intense shear deformation can be observed.

Costin *et al.* (1979) have carried out similar high-strain-rate tests on steels at different temperatures. Figure 2.15 shows experimental results at temperatures between −157 and 107 °C and two strain rates on cold-rolled 1018 steel. In almost all cases shear localization occurred by banding.

2.3.3   Effect of hydrostatic pressure on stress–strain curve in torsion

Most of the experimental work that has been concerned with the effects of superimposed hydrostatic pressure on the mechanical properties of materials has been carried out on tensile tests. Surprisingly little work has been devoted to the effects of pressure on the shear-stress–strain behaviour of materials.

Osakada *et al.* (1977) have carried out torsion tests on thin-walled steel tubes at pressures up to 4 kbar. They found that as the pressure increased the point of maximum shear stress is delayed to higher values of strain (see Fig. 2.16). This would suggest that the instability strain is influenced by

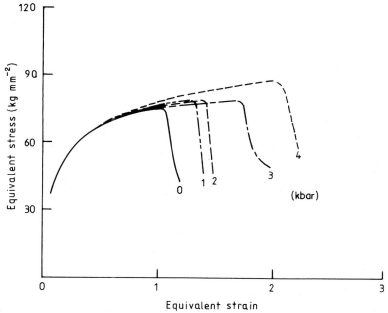

Fig. 2.16   Effect of superimposed hydrostatic pressure on torsional behaviour of a steel. (After K. Osakada *et al.*, *Bull. JSME* **20**, 1557–1565 (1977).)

geometrical effects by analogy with the effects of hydrostatic pressure on tensile tests as discussed in §2.1.6. However, this does not agree with the fact that there are no dimensional changes in torsion.

This can be rationalized by considering the microstructure of the steel. It is quite possible for cracks or microvoids to nucleate at pearlite colonies and so forth. Let us say that this occurs at some average critical strain γ_c. Then for strains larger than γ_c voids grow in size and thereby decrease the net section of the specimen undergoing shear deformation. The possibility of microstructural geometrical softening in shear will be returned to and discussed in more detail in Chapter 6.

3

Effects of Hydrostatic Stress, Temperature and Strain Rate on Ductility

Before studying the details of ductile fracture, it will be helpful to obtain a guide to the variables that affect ductility. Some of the important factors are temperature, loading conditions, hydrostatic stress and testing atmosphere. Among these effects, arguably the most significant are hydrostatic stress, temperature and strain rate.

In general, increases in compressive hydrostatic stress and temperature both favour an increase in ductility; conversely an increase in strain rate normally results in a decrease in ductility. Usually these factors and their corresponding effects are not difficult to observe. Nevertheless, knowledge of them remains incomplete and there is still controversy about certain of the experimental observations. Part of the difficulty lies with the necessity to uncouple the effects of each variable on ductility. In this book, though, we are not concerned with ambiguities and controversies about particular experimental results; rather we are concerned with the general trends of behaviour caused by hydrostatic stress, temperature and strain rate. It is hoped that a description of the general trends of behaviour will be of assistance in research and development work as well as in forming a basis to understand later chapters.

3.1 Effect of hydrostatic stress on ductility

The pioneering experimental work of von Kármán (1911) on marble and sandstone subjected to superimposed compressive hydrostatic stress revealed a valuable general concept that compressive hydrostatic stress enhances the ductility of materials. Conversely, a superimposed tensile hydrostatic stress decreases ductility and can lead to premature fracture.

von Kármán used cylindrical specimens that could be subjected to the combined action of hydrostatic stress and an independent additional axial stress. Both stresses could be varied and measured accurately. von Kármán noted considerable increases in ductility with increasing compressive hydrostatic stress for both marble and sandstone. Figure 3.1 presents the results for marble.

Subsequent to von Kármán, numerous researchers have studied the effects of hydrostatic stress on the ductility of materials, Bridgman's (1952) work being amongst the most famous. All this subsequent work confirms the general remarks made above, viz the higher the compressive hydrostatic stress the higher the observed ductility.

Brandes (1970) presented a comprehensive picture of the effects of

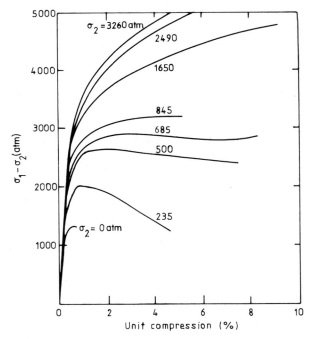

Fig. 3.1   von Kármán's axial compression measurements on marble performed at the indicated pressures. Note the remarkable increase in ductility with increasing pressure. The axial stress is σ_1 and the lateral stresses are $\sigma_2 = \sigma_3$. σ_1 was always increased above the ambient hydrostatic pressure; therefore the ordinate is $\sigma_1 - \sigma_2$.

hydrostatic stress on the mechanical properties of materials. Except where noted to the contrary, the material presented in following subsections is adopted from Brandes.

3.1.1 Ductility

The ductility of a material is here defined as the logarithm of the ratio of the original to the final cross-sectional area in a rod tensile test: $\varepsilon_f = \ln (A_0/A_f)$. A large amount of data shows that the ductility increases with increasing compressive hydrostatic stress for nearly all metals (see Fig. 3.2). An important feature shown in Fig. 3.2 is that there is an abrupt change or transition from brittle to ductile behaviour for some metals, mainly nonferrous ones. This is a type of brittle–ductile transition—below a critical pressure such metals behave in a brittle manner; however, at pressures just above the threshold a sudden increase in ductility occurs.

An explanation for this pressure-induced transition may be proposed for metals with a low rate of strain hardening (e.g. zinc). The fundamental concept is that fracture is governed by a characteristic tensile stress (which we shall discuss further in §3.1.3.). A tensile specimen tested while being subjected to a superimposed compressive hydrostatic stress requires extra stress and strain increments to offset the compressive stress. Materials with stress–strain curves that are "flat-topped" (Fig. 3.3b)—as opposed to curves exhibiting more strain hardening (Fig. 3.3a)—will show a sudden increase in ductility beyond a critical pressure.

Pressure-induced brittle–ductile transitions are also attributed to a change in slip systems in hexagonal metals. Davidson et al. (1966) also found that the transition pressure for magnesium decreases as the temperature increases. This shows once again that both increasing hydrostatic pressure and temperature increase ductility.

An increase in strain to fracture in torsion tests is also observed in most materials with increasing compressive hydrostatic stress (see §2.3.3.). This is most probably associated with the effect of superimposed pressure on the nucleation and growth of voids, which is often the mechanism of ductile fracture, this will be discussed further in Chapter 4.

3.1.2 True-stress–true-strain curve

From the load–extension curve, or equivalently the nominal stress–strain curve, it can be seen that superimposed hydrostatic pressure has little effect on the maximum load, the flow stress and the yield stress (see Fig. 3.4). The materials investigated by Pugh (see Brandes, 1970), were copper, tungsten and two different types of steel. Only when a material is subjected to very high

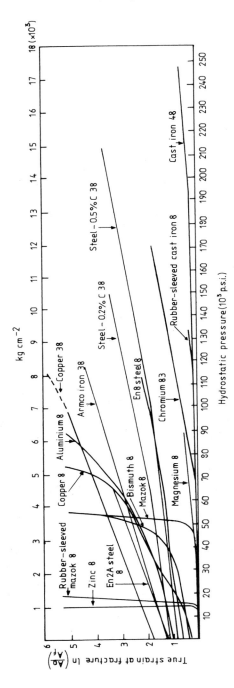

Fig. 3.2 Effect of hydrostatic pressure on the ductility of several metals and alloys. (After M. Brandes, in *The Mechanical Behaviour of Materials under Pressure* (ed. H. L. D. Pugh), 1970, with permission from Elsevier Applied Science Publishers Ltd.)

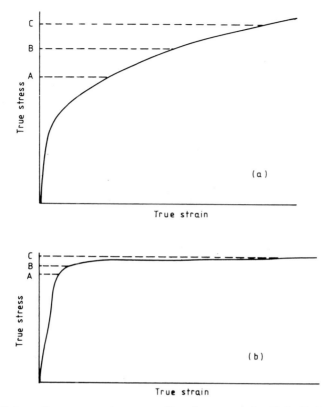

Fig. 3.3   Schematic stress–strain curves. (Based on curves by H. L. D. Pugh and D. Green, *Proc. Inst. Mech. Engrs* **179**, 415–437 (1964/65).)

compressive hydrostatic stresses is the yield stress slightly increased. The strain to fracture is a sensitive function of the hydrostatic pressure, whereas the stress levels of the stress–strain curve do not appear to be pressure-sensitive.

The true-stress–true-strain curve reveals some interesting features concerning necking; these features are of importance in understanding ductility. Simultaneous measurements of the load and the current minimum cross-section of the neck, although tedious and subject to inaccuracies (even using photographic techniques), provide valuable information. Examples of true-stress–true-strain curves for copper specimens tested under various pressures are shown in Fig. 3.5. For a number of materials, hydrostatic pressure has little effect on strain hardening, and this is shown in Fig. 3.5 for copper. Other

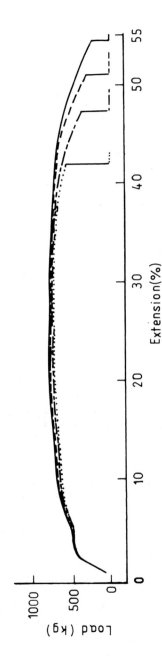

Fig. 3.4 Load-extension curves for steel 20 (0.2% C), recorded under different hydrostatic pressures: (.....), atmospheric; (———), 4000 kg cm⁻²; (- - -), 6000 kg cm⁻²; (———), 7000 kg cm⁻². (After M. Brandes, in *The Mechanical Behaviour of Materials under Pressure* (ed. H. Ll. D. Pugh), 1970, with permission from Elsevier Applied Science Publishers Ltd.)

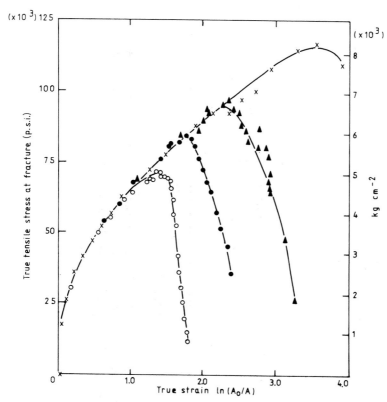

Fig. 3.5 Effect of pressure on the true stress–strain curve for annealed OFHC cooper: (○), 15 p.s.i.; (●), 11 200 p.s.i.; (△), 22 440 p.s.i.; (×), 44 800 p.s.i. (After H. L. D. Pugh, *ASTM Special Technical Publication* 374 (1964).)

important features of the results shown in Fig. 3.5 are the increase in strain to fracture with increasing hydrostatic pressure and the shape of the curves after the peak stress.

In the last sub-section we discussed the pressure-induced brittle–ductile transition by superposition of stresses. This type of description is also helpful in understanding the pressure sensitivity of the true stress.

There are no macroscopic geometrical effects that can be responsible for the decrease in stress with increasing strain after the peak stress. Therefore this softening effect with increasing strain must derive from the properties of the material. Bridgman assumed that this softening effect is not an intrinsic property of the material, but it results from insufficient homogeneity in the original material. A further possible reason is that it is quite likely that, in the region of the axis at the plane of minimum section of the necked tensile specimen, a tensile hydrostatic stress state there may initiate microvoids or cracks. If this is so then the presence of microvoids will make the real area supporting the load smaller than it appears (see Fig. 1.3).

Spretnak (1974) noticed that the difference in the strains at the maximum load ($dL = 0$) and the maximum stress ($d\sigma = 0$) increases as the hydrostatic pressure increases. From this point of view, superimposed compressive hydrostatic stress prolongs the incubation period for ductile fracture between the two kinds of instability $dL = 0$ (maximum load) and $d\sigma = 0$ (maximum true stress).

3.1.3 Fracture stress

It would be very desirable to develop a universal stress criterion for fracture under superimposed hydrostatic stress. After examining various stress combinations at fracture, Bridgman concluded that there is no such criterion. However, for a high-carbon steel, Bridgman found that the mean total hydrostatic pressure, given by

$$\Sigma = \tfrac{1}{3}(\sigma_z + 2\sigma_r + 3p)$$

remains approximately constant, where σ_z and σ_r are the true stresses at the centre of the neck and $-p$ is the hydrostatic pressure (see Fig. 3.6). Pugh found, however, for steels and copper, that the true stress at fracture is nearly proportional to the pressure, as shown in Fig. 3.7. These results have the same tendency as those shown in Fig. 3.6. Pugh came to the further conclusion that the actual stress $\sigma_z + p$ at fracture is almost independent of pressure. The only way that this conclusion can be coincident with Bridgman's observations on high-carbon steel is if $\sigma_r + p$ is constant. Interestingly, Bridgman showed that σ_r increases slowly as the pressure also increases, and at some stage $\sigma_r + p = 0$. Thus the neck of the tensile specimen may sustain a transition from tensile lateral stress ($\sigma_r + p > 0$) to a compressive lateral stress ($\sigma_r + p < 0$). This transition will cause a transition from a cup-and-cone to a shear fracture.

It is important to note that Pugh's experimental results provide a basis for assuming that fracture occurs at a characteristic true tensile stress. Based on this work, an explanation has been given for the increase in ductility with increasing pressure and the abrupt pressure-induced brittle–ductile transition.

3.1.4 Fracture modes

As discussed previously, the condition $dL = 0$ may mark a type of instability; for example, geometrical softening leads to a smaller real area to support the

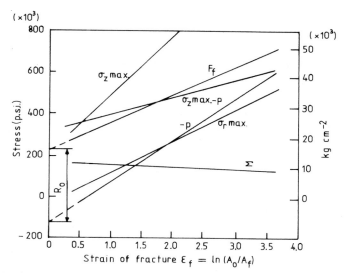

Fig. 3.6 Various possible significant stresses as a function of the strain at fracture for steel. (Based on a number of graphs by P. W. Bridgman *Studies in Large Plastic Flow and Fracture.* McGraw-Hill, New York, 1952.)

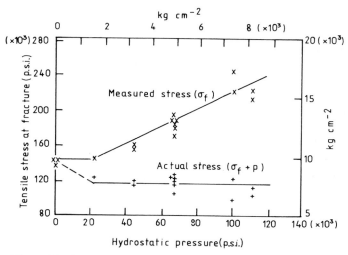

Fig. 3.7 Effect of hydrostatic pressure on the fracture stress of En 2A mild steel. (After H. L. D. Pugh, *Bulleid Memorial Lectures*, Vol. IIIB. University of Nottingham, 1965.)

load in a tensile test. Hence the effect of a compressive hydrostatic stress is to prolong the metastable condition until $d\sigma = 0$ at larger plastic strains. A suggestion by Rozovsky *et al.* (1973) to explain the above phenomenon is that high pressures diminish the effect of voids, perhaps suppressing void nucleation and growth. It follows that there may be a critical relative pressure (i.e. ratio of applied pressure to material flow stress) below which void nucleation and growth are observed. Above this critical relative pressure, void nucleation is suppressed. The concept of the critical relative pressure is valid for steels, where it has been shown to lie between 3 and 4.

More comprehensive ideas on the effects of superimposed hydrostatic stress on the modes of fracture were proposed by Davidson and Ansell (1969). They examined steels using electron fractography. For annealed 0.004%C steel fracture occurs by the nucleation, growth and coalescence of massive voids in the neck in tests carried out at atmospheric pressure. However, voids are suppressed with increasing superimposed compressive hydrostatic stress, their nucleation being suppressed completely at 15.1 kbar. The fracture profile changes from shear steps of the "hill and valley" type to a single shear plane as the pressure is increased. More examples are given by Davidson *et al.* (1966) that show the suppression of void nucleation, crack initiation by fracturing of second-phase particles and intergranular fracture with increasing compressive hydrostatic stress. After studying a number of materials, which included zinc, tungsten and water-quenched 1045 steel, Davidson *et al.* (1966) suggested that hydrostatic pressure retards those fracture modes that are susceptible to a normal tensile stress—for example cleavage, intergranular fracture and void nucleation and growth—leaving only shear instability and characteristic shear fractures. Therefore the pressure-induced brittle–ductile transition of magnesium is attributed to the suppression of intergranular fracture or void nucleation, depending on the temperature. This idea is basically an extension of von Kármán's original concept. It is not fully understood how hydrostatic pressure influences shear instability. This is very different to modes of fracture that are dependent upon the normal tensile stress. However, shear-band initiation depends upon the hydrostatic component of the stress. This will be discussed further in Chapter 6.

3.2 Transition temperatures

As the temperature is varied, metals can undergo several fracture regimes, from brittle to ductile-type fractures. At higher temperatures recrystallization and melting may be added complicating features that will influence the mode of fracture. A distinct feature of many of these changes is that they occur over a relatively narrow temperature interval; such temperature intervals are

commonly referred to as transition temperatures. In the following subsections the effects of major temperature transitions on fracture mode together
with coupling effects with other factors such as hydrostatic pressure are
introduced. The transitions in fracture mode of various metals and alloys will
be described in §3.3 in relation to fracture-mechanism maps.

3.2.1 The Ludwik–Davidenkov–Orowan hypothesis

One of the most important transitions at relatively low temperature is that
from brittle to ductile fracture. Some typical brittle–ductile transitions are
shown in Fig. 3.8. This behaviour can best be understood by the so-called
Ludwik–Davidenkov–Orowan hypothesis as described by Cottrell (1964).
The hypothesis consists of two fundamental assumptions. First, brittle
fracture may become possible when the yield stress exceeds a critical value as
the temperature is decreased. Secondly, the brittle-fracture stress is assumed
not to be as sensitive to temperature as the yield stress. This has an interesting
parallel in the insensitivity of the brittle-fracture stress to hydrostatic
pressure. Here, for simplicity, brittle fracture and plastic deformation are
examined as completely separate processes. This is shown in Fig. 3.9. Simply
$F \gtrless Y$ characterizes ductile and brittle fractures respectively, and the
temperature at which $F = Y$ is the transition temperature for brittle to ductile
behaviour. This hypothesis is physically reasonable, because to some extent
the mechanism of brittle fracture, that is cleavage, is a mechanism concerned
with the breaking of interatomic bonds. This mode of fracture is not very

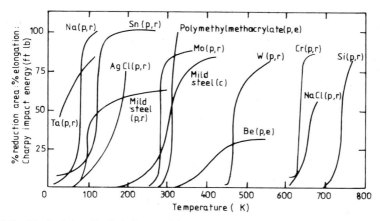

Fig. 3.8 Typical ductile–brittle transitions: p, plain tension; r, reduction of area; e,
elongation; c, Charpy V-notch impact energy. (After A. H. Cottrell, *The Mechanical
Properties of Matter*. Copyright 1964 John Wiley and Sons, London.)

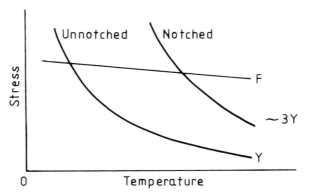

Fig. 3.9 Interpretation of plain-bar and notched-bar ductile–brittle transition temperatures. (After A. H. Cottrell, *The Mechanical Properties of Matter.* Copyright 1964 John Wiley and Sons, London.)

sensitive to temperature, whereas ductile fracture must be preceded by plastic flow and often void nucleation and growth, for which thermal activation mechanisms are usually assumed to be essential.

This hypothesis can also be used to understand how the brittle–ductile transition can be lowered by superimposed compressive hydrostatic stress, as shown in Fig. 3.10 for zinc. We have already observed that increases in the hydrostatic pressure and temperature increase ductility. If a brittle specimen is subjected to a hydrostatic pressure the specimen will become ductile and the transition temperature is decreased. Although this explanation appears to be valid, it should be noted that an increase in transition temperature has not been observed for any materials when subjected to hydrostatic tension.

3.2.2 Brittle–ductile fracture transition and impact tests

As we have seen above, not only do hydrostatic stress and temperature affect ductility, but ductility is also sensitive to applied strain rate. Furthermore, the shape and size of members are also influential; that is, there is a strong size effect. Also, notches and flaws can promote brittle fracture. The most important conclusion is that in-service conditions are almost invariably more severe than those observed in the laboratory. Tests have been developed to assess the notch sensitivity of materials subjected to impact, particularly steels. Common tests are the Charpy, Izod, Robertson and slow-bend tests of a notched sample. In Fig. 3.8 typical ductile–brittle fracture transitions are shown for various materials. Here a ductile material is one that absorbs a significant amount of energy before it fractures. Also, equivalently, the

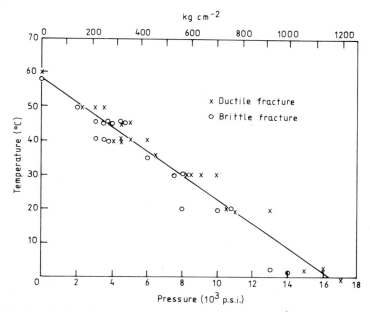

Fig. 3.10   Brittle–ductile transition temperature for zinc as a function of hydrostatic pressure. (After H. L. D. Pugh, *Bulleid Memorial Lectures*, Vol. IIIB. University of Nottingham, 1965.)

reduction in area or elongation in uniaxial tensile tests can be plotted. From Fig. 3.8 it can be seen that a notched steel specimen has a higher transition temperature than an unnotched specimen. The differences in behaviour of other metals are also shown. The refractory metals such as niobium and tantalum have low transition temperatures. The general trend appears to be that the transition temperature is lower the higher the hardness.

There are several practical definitions of transition temperature. The nil-ductility temperature (NDT) signifies the point on the energy–temperature curve at which the curve begins to rise. At the fracture-appearance transition temperature (FATT), the appearance of the fracture surface is 50% crystalline (brittle) and 50% fibrous (ductile). Additionally, there are two temperatures that are commonly defined in the slow notch-bend test. These are T_{GY}, above which the specimen plastically yields, and T_W, which is the temperature associated with the yielding of the gross section of the specimen before fracture and the spread of plastic "wings" from the top face of the specimen (Knott, 1973). Schematic relationships between these temperatures, energies and loads are shown in Fig. 3.11.

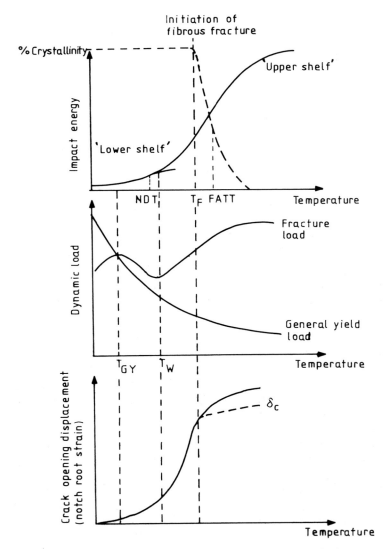

Fig. 3.11 Schematic relationship between quantities involved in impact testing. (After J. F. Knott, *Fundamentals of Fracture Mechanics*. Butterworths, London, 1973.)

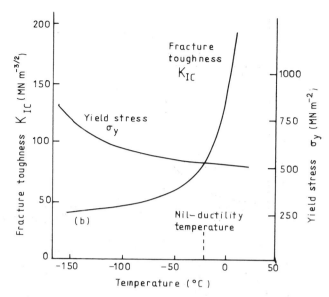

Fig. 3.12 Variation of K_{IC} and σ_y with temperature for a low-alloy structure steel A533B. (After E. G. Wessel, *Practical Fracture Mechanics for Structural Steel*, 1960, Paper H, UKAEA/Chapman and Hall, London.)

One very important feature of these curves is that they are derived from different types of tests and imply different strain rates, specimen sizes and so forth, according to which particular service conditions are being simulated.

The plane-strain fracture toughness K_{IC} also shows a transition with change in temperature, especially for relatively low-strength structural steels. The main obstacle in testing such materials is the physical size of specimens required to ensure plane-strain conditions. Figure 3.12 shows the relationship between the fracture toughness and temperature for A533 Grade B Class 1 steel. Above approximately the NDT the fracture toughness increases rapidly with increasing temperature. On the other hand, the fracture toughness of nonferrous alloys and high-strength steels, like maraging steels, vary little with change in temperature.

3.2.3 Transitions at elevated temperatures

It was noted in Chapter 1 that as the temperature is increased ductile fracture associated with voids can be governed by plastic flow, creep and/or diffusion. As an example of these and other related transitions we will consider the case of iron as an example, after Fields *et al.* (1980).

Below $0.1T_M$ (200 K), where T_M is the melting point, iron fails by cleavage. Both precracked and uncracked specimens fail by cleavage, although cleavage may be nucleated by slip-induced cracking of carbides at this temperature.

At temperatures above $0.1T_M$ the fracture surface becomes 100% fibrous, exhibiting ductile fracture. Voids nucleate and grow by plastic flow, and finally a cup-and-cone fracture ends the process.

Above about $0.3T_M$ iron begins to creep, and strain-rate sensitivity becomes the major factor governing flow, stabilizing it. Both intergranular and transgranular fractures may occur, depending both on the value of the stress and the time at the stress.

Above $0.5T_M$ phase transformations and recrystallization may occur. For iron the Curie temperature is at $0.57T_M$, and the body-centred cubic to face-centred cubic phase transformation $(\alpha \rightarrow \gamma)$ occurs at $0.65T_M$. Above these temperatures recrystallization occurs with further plastic deformation. Finally, at temperatures above about $0.7T_M$ rupture occurs, which is caused here by dynamic recrystallization.

Although the above outline of transitions has concentrated on iron, there are similar patterns for other metals.

Other factors that may become important as the temperature is raised are the rate of strain hardening and the strain-rate sensitivity. As the temperature is increased, the rate of strain hardening decreases, also the strain-rate sensitivity can become marked at elevated temperatures, as shown in creep. Both of these factors influence localized flow, such as the region or ligament between voids; therefore these factors must have an important influence on ductility.

3.2.4 Recrystallization

Recrystallization occurs above a certain temperature, called the recrystallization temperature T_r, which is not so well-defined as the melting point. Generally, the recrystallization temperature ranges between 0.4 to $0.55T_M$, where T_M is measured in Kelvin.

Six main variables influence recrystallization behaviour. These are (1) the amount of prior deformation, (2) the temperature, (3) the time, (4) the initial grain size, (5) the composition and (6) the amount of prior recovery. It is clear that because the recrystallization temperature is dependent upon the above six variables it is not fixed in the same way as the melting point.

At constant temperature, recrystallization is found to consist of three consecutive stages. The first stage is nucleation, which defines a delay or nucleation time. After this time new grains begin to grow and the growth rate is almost constant. Finally the growth rate decreases to zero because of interaction of the new grains.

In the light of the above observations, it is clear that recrystallization is a thermally activated process. Thus intense areas of deformation such as deformation bands or interfaces tend to be nucleation centres for new grains.

Above the recrystallization temperature the ratio of the dynamic to the quasistatic flow stresses becomes very sensitive to temperature (see Fig. 3.13). At temperatures below the recrystallization temperature the ratio is between 1 and 2, whereas above this temperature the ratio increases dramatically. This feature is particularly important in fast upset forging of hot specimens, where the relative strength can be higher than that for cold deformation. The rate of softening offsets the rate of strain hardening. Which one of these effects dominates depends upon whether deformation is occurring above or below T_r.

Recrystallization can occur during plastic deformation; this is called dynamic recrystallization. Dynamic recrystallization allows a rapid relief of the stored energy of cold work and commonly leads to rupture in most metals.

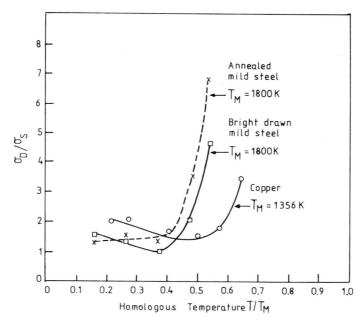

Fig. 3.13 Variation of dynamic to static mean yield stress ratio with homologous temperature. (After W. Johnson *Impact Strength of Materials*. Arnold, London, 1972.)

3.3   Fracture-mechanism maps

Fracture-mechanism maps have already been mentioned in relation to the
outline of fracture modes in Chapter 1. Here we shall use fracture-mechanism
maps to help understand the effects of temperature, stress and loading period
on ductile fracture. Fracture mechanism maps are usually drawn for uniaxial
tensile tests of rod specimens.

There are several types of fracture map, but they all have one common
feature; that is, to identify the distinct fracture modes with the governing
variables, for example stress and temperature, as coordinates.

Lindholm (1974) used effective stress and effective strain as axes, while
Wray (1969) used strain rate and temperature. Clearly, different axes will
emphasize the roles of different variables on fracture.

Ashby *et al.* (1979) have collected an enormous quantity of data on fracture
modes of different metals as functions of temperature, stress, time and so
forth, and their fracture-mechanism maps are in consequence more comprehensive than those of previous workers. Accordingly, most of the following
description comes from research of Ashby and his co-workers (see Frost
and Ashby (1982) for an up-to-date compendium of fracture mechanism
maps).

3.3.1   Basic features of fracture-mechanism maps and their construction

There are two related types of fracture maps. One is the stress–temperature
map, which plots the normalized tensile stress σ_n/E versus the homologous
temperatures T/T_M, where E is Young's modulus, T_M is the melting point, and
the temperatures are in Kelvin. The other type of fracture-mechanism map is
plotted on axes of stress and time to fracture t_f.

The construction of fracture mechanism maps is based solely on experimental data. The relevant testing parameters are the tensile stress σ_n,
temperature, time to fracture t_f and the fracture strain ε_f and must be
collected. Depending on the experimental data and fractographic observations as well as macroscopic and scanning electron microscope examinations, each point on the map is plotted together with symbols that indicate
mode of fracture. When a large number of points have been plotted, a fracture
mechanism map emerges. The types of fracture observed can be cleavage
types 1, 2 and 3, brittle intergranular fracture (BIF) 1, 2 and 3, ductile
fracture, intergranular and transgranular creep fractures, rupture and
dynamic fracture. Also it is possible to draw on the map contours of constant
time to fracture and further plot the normalized tensile stress versus t_f. An
illustrative example has already been given in Fig. 1.1. Figure 3.14 is a plot of
σ_n/E versus t_f for 304 stainless steel.

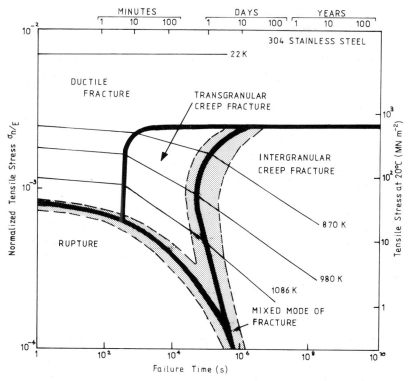

Fig. 3.14 A fracture-mechanism map of the second type for low-boron, homogenized and quenched 304 stainless steel. (After R. J. Fields *et al.*, *Met. Trans.* **11A**, 333–347 (1980).)

Some very general remarks on fracture modes will be helpful here. At low temperatures cleavage controls fracture. Cleavage 1 corresponds to a low-stress fracture due to pre-existing flaws and cleavage 2 is used to differentiate between cleavage 1 cracks and those that are formed by slip or twinning. Cleavage 3 corresponds to fracture associated with significantly more plasticity. Cleavage 3 occurs at a higher temperature than cleavage 2. At higher temperatures than the cleavage 3 zone, ductile fracture clearly occurs. At lower normalized tensile stresses than the zone over which this ductile fracture occurs, transgranular and intergranular creep fractures occur. From Fig. 3.14 this implies that at a constant temperature, say for example, 870 K, ductile fracture occurs in the area of highest normalized tensile stress and the shortest times, typically less than 10 min. At 870 K, if the specimen is subjected to a lower loading then the material will not fracture unless held under load for some time, say a few hours; under these conditions transgranular creep fracture will occur. Rupture occurs in the region of lowest normalized stress and failure times.

In metalworking and similar processes in which the loading time may be relatively short, from Fig. 3.14 ductile fracture governed by plastic flow predominates as the important failure mechanism. Dynamic fractures associated with stress-wave effects are the highest stress-level fractures over the entire temperature range according to Ashby *et al.* (1979).

Unfortunately, it is very difficult to draw a corresponding zone on the stress–time map; it is nevertheless possible to visualize the dynamic fracture zone as being placed at the top left-hand corner of Fig. 3.14, for example.

3.3.2 Fracture-mechanism maps for different groups of metals

For face-centred cubic metals and alloys (e.g. nickel, silver, copper, aluminium and lead) there are four main fracture modes: ductile, inter- and

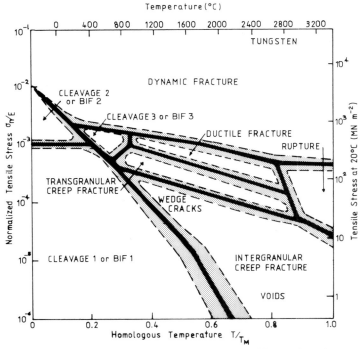

Fig. 3.15 A fracture-mechanism map for tungsten tested in tension; the map shows three modes of cleavage at low temperature, and other fields resemble those of f.c.c. metals. (After C. Gandhi and M. F. Ashby, *Acta Metall.* **27**, 1565–1605 (1979).)

transgranular fractures and rupture. For nickel creep failures occur at $0.3T_M$, and these are transgranular for times of less than 20 min. Longer loading times lead to intergranular creep fracture at lower stresses. Recrystallization occurs at about $0.6T_M$. Above this temperature rapid tensile tests result in rupture together with a fine-grain structure.

For alloys, a stable dispersion suppresses dynamic recrystallization, and both dispersions and solid solutions raise the overall stress level of the fracture favouring intergranular creep fracture. These appear to be common trends for most face-centred cubic metals and alloys.

Body-centred cubic metals like tungsten, molybdenum, tantalum and chromium include all fracture modes (see Fig. 3.15). It is clear from this example that there are differences from face-centred cubic metals like nickel. Tungsten exhibits cleavage at low temperatures, but either fractures in a ductile manner or by creep above $0.3T_M$. Above $0.8T_M$ rupture occurs. These transitions also occur for hexagonal metals and alloys.

Pure iron has all possible fracture modes, and also there are phase transformations involved. The fracture transitions have already been outlined in §3.2.3. For ductile failure voids nucleate at inclusions and/or grain boundaries and grow by plastic flow; finally a specimen will exhibit a cup-and-cone fracture.

On the other hand, austenitic stainless steels (18% chromium and 8% nickel) only exhibit ductile fracture, trans- and intergranular fractures and rupture. However, certain corrosive atmospheres may modify the possible fracture modes.

3.3.3 Changes in fracture-mechanism maps

One general conclusion that can be drawn from the study of fracture-mechanism maps for metals of the same crystal structure is that they all show similar patterns. For example, nickel and tungsten typify face-centred cubic and refractory body-centred cubic metals respectively. Understanding the different classes of fracture-mechanism maps is informative. For example, alloying and dispersion hardening raise the general stress levels of the boundaries on the maps. By this, rupture may be suppressed, while on the other hand the field of intergranular creep fracture will expand.

To obtain a general view of map alterations, Gandhi and Ashby (1979) examined the effects of different bonding as shown in Fig. 3.16. According to them, there are five groups of solids: face-centred cubic metals, body-centred cubic metals, alkali halides, refractory oxides and covalently bonded solids.

In metals with one of the three common crystalline structures all brittle and ductile fracture modes are possible, except for cleavage in face-centred cubic metals and stainless steels. It is interesting to note that the cleavage 1

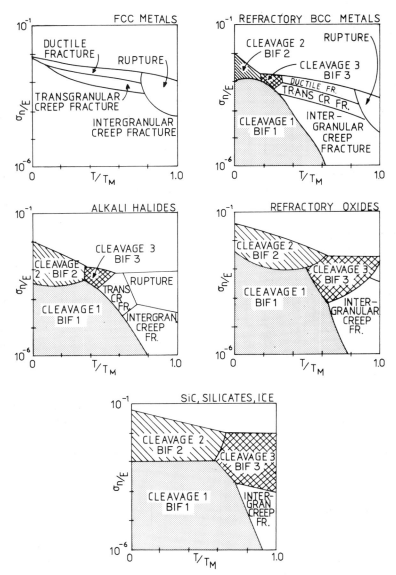

Fig. 3.16 The progressive spread of fracture-mechanism fields, from left to right, as the bonding changes from metallic to ionic and covalent. (After C. Gandhi and M. F. Ashby, *Acta Metall.* **27**, 1565–1605 (1979).)

field gradually expands as the bonding changes from metallic to ionic and covalent. Also, areas of rupture are absent for refractory oxides and silicates.

From the trends shown in Fig. 3.16, we may suppose that the fracture behaviour of solids depends on several microscopic and submicroscopic phenomena, for example atomic bonding, crystalline structure (assemblies of atoms), grains and their boundaries, inclusions and finally pre-existing flaws and cracks. The complex question is how to combine all these important variables to predict the various levels of toughness and ductility observed.

3.3.4 Criteria for brittle and ductile fractures

The prerequisite for ductile fracture is plastic deformation the mechanism of which is plastic shear or slip. For materials with strong bonding, for example, ionic or covalent bonds, there is a high lattice resistance against slip. In these cases it appears that cleavage dominates fracture. It is clear that if we know the resistance to plastic shear and the cohesive strength of the material then these can, in principle, be developed into criteria for ductile and brittle fracture. But the real problem involved here is to identify these two quantities.

One approach is to consider the ratio G/K, where G and K are the shear and bulk moduli respectively (see Table 3.1). Obviously these two quantities express some types of resistance to small deformations. The second criterion shown in the table is $G\Omega^{1/3}/\gamma$, where γ is the surface energy and Ω the atomic volume. γ is used here as some measure of the cohesive strength. Criteria 3 and 4 use the yield strength (extrapolated back to the absolute zero of temperature) as the resistance against plastic shear. Both of these latter criteria grade materials more accurately than the former two. It is obvious that if we wish to grade materials according to their mode of fracture, brittle or ductile, then it is of crucial importance to identify the relevant quantities accurately. For ductile fracture large plastic shears are necessary, and although the shear modulus G is related to plastic shear, the yield stress σ_y is more closely related to fracture mode.

3.4 High-strain-rate effects and spallation

High-strain-rate loading usually produces cracking, because it increases the yield stress and shrinks the range of plastic deformation, favouring fracture. Thus it is important to try and understand dynamic-loading phenomena, because if it occurs in service the effect may be catastrophic. A further aspect of high strain rates is that they are accompanied by heat produced by the plastic deformation itself. This temperature rise is often quite localized and induces thermal softening and a greater strain-rate dependence. In particular,

Table 3.1

Criteria for cleavage (after C. Gandhi and M. F. Ashby, *Acta Metall.* **27**, 1565–1605 (1979))

Material	G (GN m^{-2})	K (GN m^{-2})	Ω ($\times 10^{-29}$ m^3)	γ (J m^{-2})	T_M(K)	σ_y (GN m^{-2})	Criterion 1: G/K	Criterion 2: $G\Omega^{1/3}/\gamma$	Criterion 3: $\sigma_y\Omega/kT_M$	Criterion 4: σ_y/K ($\times 10^{-2}$)
f.c.c. metals										
Pb	10	47	3.0	0.6	600	0.07	0.21	5.2	0.25	0.15
Ag	30	104	1.7	1.6	1234	0.3	0.29	4.8	0.3	0.29
Cu	45	141	1.2	2.2	1356	0.4	0.32 } Mean	4.7 } Mean	0.25 } Mean	0.28 } Mean
Al	27	82	1.7	1.1	933	0.25	0.33 } 0.3	6.3 } 5.7	0.33 } 0.3	0.3 } 0.3
Ni	86	180	1.1	2.5	1726	0.73	0.48	7.7	0.34	0.41
Ir	220	360	1.4	3.6	2716	3.6	0.61	15	1.4	1.0
b.c.c. metals										
Fe	79	160	1.2	2.7	1810	1.1	0.49	6.7	0.52	0.69
Nb	39	168	1.8	3.2	2741	1.4	0.23	3.2	0.67	0.83
Ta	70	192	1.8	3.4	3271	1.5	0.36 } Mean	5.4 } Mean	0.61 } Mean	0.78 } Mean
Mo	126	265	1.5	2.9	2883	1.4	0.48 } 0.5	10.7 } 7.5	0.53 } 0.7	0.53 } 0.9
W	160	305	1.6	4.2	3683	3.8	0.52	9.6	1.2	1.3
Cr	120	160	1.2	2.9	2148	2.0	0.75	9.5	0.8	1.3

h.c.p. metals

Mg	18	36	2.3	—	923	0.69	0.5	—	1.3	1.9
Be	150	116	0.8	0.9	1588	2.6	1.29	33	0.96	2.2
Re	186	363	1.5	—	3453	3.4	0.51	—	1.0	0.94
Mean							0.8		1.1	1.7

Alkali halides

NaCl	15	25	4.5	0.28	1073	0.47	0.6	19	1.4	1.9
KCl	10	19	6.3	0.16	1063	0.36	0.53	25	1.5	1.9
LiF	49	68	1.6	0.34	1143	3.0	0.72	36	3.1	4.4
CaF$_2$	42	91	—	—	1691	2.1	0.46	—	—	2.3
Mean							0.6	27	2.0	2.6

Oxides

MgO	129	154	1.9	1.2	3073	7.5	0.84	29	3.4	4.9
Al$_2$O$_3$	160	247	1.4	1.0	2323	13.5	0.65	39	5.9	5.5
BeO	163	236	1.4	—	2701	6.1	0.69	—	2.3	2.6
UO$_2$	85	212	4.1	—	3151	3.6	0.40	—	3.4	1.7
Mean							0.6	34	3.8	4.0

Covalent and H-bonded solids

SiC	192	227	1.04	—	3110	18	0.85	—	4.4	7.9
Si$_3$N$_4$	122	194	—	—	2173	10	0.63	—	—	5.2
H$_2$O	3.5	8.5	3.3	0.1	273	0.55	0.41	11	4.7	6.5
Ge	55	75	2.3	—	1210	5.5	0.73	—	7.6	7.3
Si	67	99	2.0	—	1683	7.8	0.68	—	6.7	7.9
Mean							0.7		5.9	7.0

special localizations of plastic flow may occur owing to these thermal effects, and in turn the localizations will affect the onset and development of fracture.

A feature of some dynamic-loading tests is spallation, which is not unrelated to ductile fracture. Spallation is a typical fracture induced by tensile-stress waves. Many of the processes that are of importance in ductile fracture, such as void nucleation, growth and coalescence, are involved in spallation. As the loading time is short and is controllable in many dynamic tests, the progress of these various processes can be traced to some extent. From stopping a number of tests part of the way through the fracture process, it is possible to investigate how ductile fracture evolves under various loading rates. Therefore the objectives of this are to reveal the effects of elevated strain rate on ductile fracture and to explore spallation as a tool for understanding ductile fracture.

3.4.1 Yield and flow stresses at elevated strain rates

In general it is found that as the strain rate is increased the yield stress increases. For example, the ratio of the dynamic yield stress for pure iron measured at a strain rate of about 10^2 s^{-1} to that measured at quasistatic rates is about 1.5.

Another aspect of yielding under dynamic loading is the yield-point delay. Pure iron can exhibit a delay time of up to 100 s. In general, the higher the stress, the shorter the delay time.

Strain rate has a similar effect on the flow stress as on the yield stress. In the range of strain rate between about 10^{-3} s^{-1} and 10^3 s^{-1} the flow stress is approximately proportional to the logarithm of the strain rate. At strain rates higher than about 10^3 s^{-1} the flow stress usually increases dramatically with strain rate, and there is often a linear relation between stress and strain rate. Many aluminium alloys and high-strength steels are not as sensitive to strain rate as iron; also, close-packed hexagonal metals and alloys do not appear to be very strain-rate sensitive, as described by Campbell (1973). Valuable compilations of data that illustrate the effects of strain rate and temperature on the yield and flow stresses of various metals and alloys have been made by Suzuki et al. (1968), Lindholm and Bessey (1969) and Eleiche (1972).

The increase of flow stress with strain rate is temperature-dependent. The higher the temperature, the more sensitive to temperature is the increase in the flow stress due to an increase in strain rate. Of course, recrystallization complicates this behaviour. Also, the yield-point delay at fixed stress decreases with increasing temperature, as shown in Fig. 3.17.

All these effects imply that the processes preceding ductile fracture are strongly rate-dependent.

In general, materials will tend to be more brittle at higher strain rates

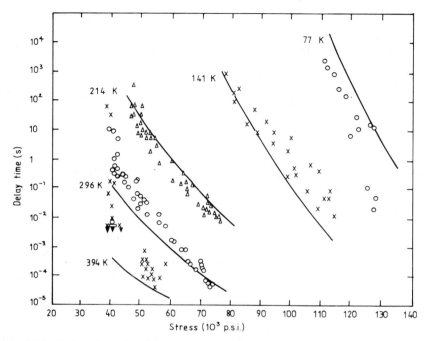

Fig. 3.17 Delay times for yield in steel at various stresses and temperatures. (After A. H. Cottrell, *Properties of Materials at High Rates of Strain*. Inst. Mech. Engrs, London, 1957.)

because the higher stress level would make the plastic region surrounding a stress concentration smaller.

3.4.2 Effect of high strain rate on ductile fracture

Zener and Hollomon (1944) proposed a parameter given by

$$Z = \dot{\varepsilon} \exp\left(U/RT\right) \tag{3.1}$$

where $\dot{\varepsilon}$ is the strain rate, U is an activation energy and R is the universal gas constant. This parameter is most commonly used to describe the combined effects of strain rate and temperature. At a given stress and strain Z is a constant. MacGregor and Fisher (1946) used an alternative concept of velocity-modified temperature T_v given by

$$T_V = T\{1 - A \ln\left(\dot{\varepsilon}_p/\dot{\varepsilon}_1\right)\}, \tag{3.2}$$

where A and $\dot{\varepsilon}_1$ are constants related to the activation process for plastic flow. Further, at a given strain the flow stress is a unique function of T_v; see Harding (1981) for a detailed discussion of mechanical equations of state.

Lindholm (1974) used a Zener–Hollomon-like thermally activated parameter given by $T^* = T \ln (A/\dot{\varepsilon})$. Here $\dot{\varepsilon}$ is the effective strain rate and A is a constant. Figure 3.18 shows the dependence of deformation and fracture upon the Zener–Hollomon parameter. The boundary between elastic and stable plastic deformation is the yield surface, whereas the boundary between stable and unstable plastic deformation is the maximum in the engineering stress, after which there is either fracture or rupture. The effects of temperature on the behaviour are the inverse of those of strain rate.

A high strain rate causes various transition temperatures to occur prematurely, such as T_{GY} and T_W. According to the Ludwik–Davidenkov–Orowan hypothesis, the brittle-fracture stress is an intrinsic material constant and is not very sensitive to temperature variations. Also, from the Zener–Hollomon parameter the brittle fracture stress is insensitive to strain rate. However, as mentioned above, increasing the strain rate usually increases the yield stress, which in turn make cleavage more possible. This is an explanation for changes in the transition temperature.

An elevated strain rate also has a softening effect caused by the heat resulting from the plastic work. The effect is accentuated in some localized

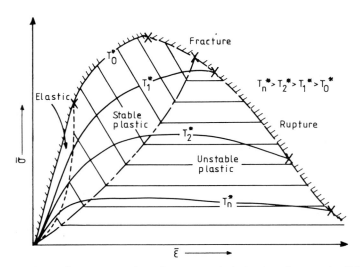

Fig. 3.18 Schematic representation of stress–strain behaviour for metals. (After U. S. Lindholm, in *Mechanical Properties at High Rates of Strain* (ed. J. Harding). Conf. Ser. No. 21, Inst. Phys., London, 1974.)

deformation zones, as discussed by Rogers (1979). In a localized deformation zone the plastic strain may be as high as 100. This would lead to a very large plastic work of 10^3 MJ m^{-3} if $\sigma_y \sim 10^7$ N m^{-2}. If deformation occurs under adiabatic conditions there would be a temperature rise of about 10^3 K in the band. Under these conditions of high strain rate and plastic strain, materials tend to behave viscously, or, in other words, exhibit marked strain-rate sensitivity. This viscous behaviour tends to stabilize plastic flow, although at the same time high temperatures soften materials. Possible complicated effects of strain rate combined with temperature are discussed in the next subsection.

3.4.3 Strain-rate effects on fracture toughness

A number of methods have been developed to measure the fracture toughness K_{IC} under dynamic loading conditions, often referred to as K_{Id}. Figure 3.19 shows the results from one such method, plotting K_{IC} versus the strain rate. A high strain rate enhances brittleness; this is the opposite effect to that of temperature. However, there is an inverse rate sensitivity associated with dynamic ageing.

The general trend seems to be that the yield stress and toughness are both sensitive to strain rate. As the strain rate increases, the yield stress increases,

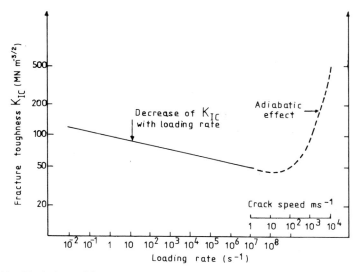

Fig. 3.19 Variation of fracture toughness with loading rate in a semi-killed steel. (After G. Irwin and A. A. Wells, *Met. Rev.* **10**, 223–270 (1965).)

but the fracture toughness decreases. On the other hand, for high-strength steel the yield stress and the fracture toughness change little with increasing strain rate; similar results have been found for maraging steel.

Referring to Fig. 3.19, which is for a semikilled steel, the decrease in fracture toughness is about 10% for each decade of strain-rate increase. However, at very high loading rates, like that ahead of a running crack, the fracture toughness may increase rapidly. Also, at the head of a running crack the heat generated cannot be dissipated in the bulk of the test piece. Thus it can be seen that both the temperature and the fracture toughness increase at very high strain rates. A related effect occurs when a running crack slows down in a normally ductile fracture; glide appears from the crack tip—the crack becoming ductile.

A further very important effect is that the energy required for ductile fracture is closely related to the strain rate. This has been shown by computer simulations made by Curran et al. (1977) for Armco iron. For high strain rates a higher energy is required to fracture metals. The interrelationships between the distribution of energy and the strain rate demonstrates that it is too simplistic to adopt the fracture energy as a material constant. It follows from this that the rate processes involved in ductile fracture are of crucial importance.

3.4.4 Spallation and rate processes

A tensile-stress wave, which is usually created by the reflection of a compressive-stress wave at a free surface, can produce scabbing near the surface. This type of failure is called spallation, and it is a strongly time-dependent failure mode. This is illustrated by Fig. 3.20. The impulsive load is

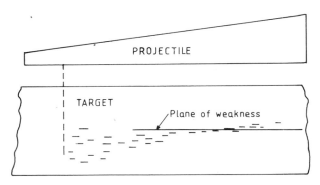

Fig. 3.20 A cross-sectional sketch of a spall in an aluminium alloy. (After Butcher *et al., J of AIAA* **2**, 977–990 (1964).)

produced by a tapered flying plate. According to the mechanics of stress-wave propagation, the thickness of the plate dominates the duration of the tensile-stress wave in the target. It is clear from Fig. 3.20 that there is a lower limit on the thickness of the flying plate beyond which no spallation can occur.

Ductile spallation is characterized by the nucleation and growth of numerous microcracks or voids, as has been shown by Seaman *et al.* (1976). As we have observed previously, the loading period is controllable in spallation tests. Hence these tests allow the examination of the development of failure from recovered specimens. From such studies, it is observed that ductile fracture comprises numerous strain-rate-sensitive processes. This is so even in fracture produced by quasistatic deformation, but is more easily observed in elevated-strain-rate tests. For both quasistatic and dynamic loadings, the microcracks or voids can nucleate and grow at a number of points within the specimen. Therefore if the area in which the microscopic rate-dependent processes develop is large in comparison with the specimen size then these microscopic processes should be taken into account. In this way, spallation is an important method for studying ductile fracture.

Curran *et al.* (1977) have used dynamic techniques to study microscopic mechanisms of ductile fracture. A gas gun is used to fire a thin flying plate at a flat target specimen. By sectioning the recovered specimen, it is possible to investigate the incipient stages of nucleation and growth in voids. It is also possible to take into account distribution of void nucleation sites and sizes using statistics. However, although this general approach is very promising for the study of ductile fracture, much more work is required to develop links between the experimental data and microscopic models.

4

Void Nucleation, Growth and Coalescence

A number of variables can influence the ductility of a metal. We have already seen that factors such as superimposed hydrostatic pressure, temperature and strain rate can change ductility profoundly. In this chapter we are concerned with the influence of the microstructure of the metal on ductility and, in particular, the role of inclusions, second-phase particles and grain boundaries as crack or void nucleation sites.

Ductile fracture commonly occurs progressively, with void or crack nucleation at inclusions or particles, the growth of these voids with increasing plastic strain, and finally the coalescence of the voids. Thus it follows that the presence of particles in the microstructure can markedly influence ductility.

In a commercially pure metal there will always be a population of inclusions or particles, for example oxide particles in copper or manganese sulphide inclusions in steel. The volume fraction, spatial and size distribution, shape and relative strength of the particles with respect to the matrix, no matter whether they are nonmetallic inclusions or second-phase particles, may all be of importance in relation to the ultimate ductility.

The ease of void initiation will depend in part on the nature of the bonding between the particle and the matrix. If the bonding is negligible or nonexistent (e.g. manganese sulphide inclusions in steel) then voids can nucleate at relatively small plastic strains. The presence of an unbonded particle in an otherwise uniform matrix will induce internal localized stresses. These stresses arise from differences in the coefficients of thermal expansion of the particle and matrix. Under some conditions these internal stresses can be very high and may suppress void initiation until somewhat higher strains. For particles which are chemically bonded to, or wetted by, the matrix (e.g. copper oxide particles embedded in copper) high localized stresses are required for void nucleation. The stress state generated by the presence of a particle was calculated for an elastically stressed matrix by Goodier (1933). If the particle is weaker than the matrix then high stresses will occur at the equator, which is similar to a hole. On the other hand, if the particle is stiffer than the matrix then elastic stress concentrations occur at the particle poles. However, ductile fracture involves plastic deformation in the matrix. In the more usual cases of a plastically deforming matrix and either a harder elastic particle or a softer plastic one the situation is far more complex than that envisaged by Goodier in his elasticity calculations. Even so, Goodier's calculations can give an indication of potentially dangerous positions in the region of a particle.

When a metal is plastically deformed, cracks or voids are nucleated either at the inclusion/matrix interface by decohesion or by particle cracking. Thus it follows that the greater the volume fraction of inclusions and second-phase particles the smaller is the ductility. This was shown to be the case in the well-known work of Edelson and Baldwin (1962) (see Fig. 4.1). For a series of

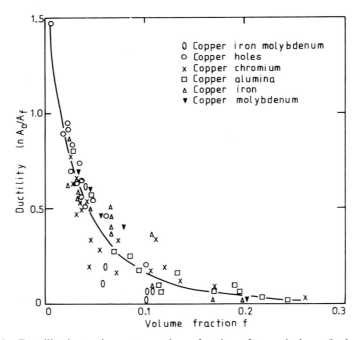

Fig. 4.1 Ductility in tension versus volume fraction of second phases/inclusions for various copper alloys. (After Edelson, B. I., Baldwin, W. M., *Trans. ASM*, **55**, American Society for Metals, Metals Park, OH 44073, 1962, pgs. 230–250.)

copper-based alloys there is an almost exponential decrease of tensile ductility with increase in particle volume fraction. These copper-based alloys were produced by powder-metallurgy techniques and the particles were not strongly bonded to the matrices. In the work of Pickering (1978) it was demonstrated that the volume fraction of second-phase particles also affects the ductility of steel significantly (see Fig. 4.2).

Plastic deformation may often cause large changes in the shape of inclusions. One of the most often quoted examples is the effect of hot working on the aspect ratio of initially spherical manganese sulphide inclusions. It has been found that hot working can increase the aspect ratio by a factor of ten or more (see Sekiguchi and Osakada, 1983), with the inclusions strung out in the direction of working. This effect can produce a significant anisotropy in ductility. The difference between the tensile ductilities of a steel in the transverse and longitudinal directions can be quite large, with the transverse ductility being the smaller, as shown in Fig. 4.3. Clearly, any method of inclusion-shape control will in turn control ductility. Relatively small plastic

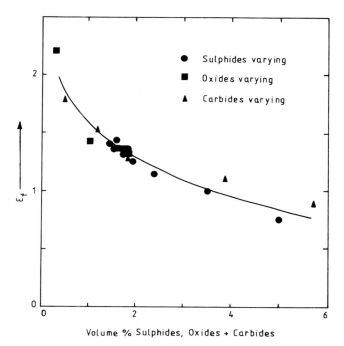

Fig. 4.2 Effect of total volume fraction of second phase particles on the tensile ductility of steel. (After F. B. Pickering, *Physical Metallurgy and the Design of Steels*, 1978, with permission from Elsevier Applied Science Publishers Ltd.)

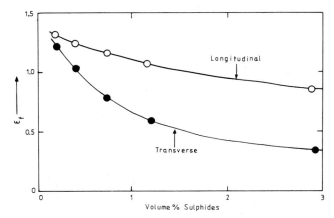

Fig. 4.3 Effect of elongation of sulphide particles on the tensile ductility of steel. (Adapted from F. B. Pickering, *Physical Metallurgy and the Design of Steels*, 1978, with permission from Elsevier Applied Science Publishers Ltd.)

strains of less than 5% are often sufficient to nucleate microcracks, for example in pearlite colonies in steels. But these microcracks may well be metastable, and a significantly larger strain may be required to make the microvoid propagate.

4.1 Fracture in single crystals

The classical work on the yielding and plastic flow of single crystals is that of Schmid and Boas (1950). Much of the initial work on the determination of the critical resolved shear stresses for numerous single crystals was done by these authors. However, little work is reported in this reference on modes of fracture in single crystals. But it is of great interest in the present context to describe ductile fractures in crystals in relation to fracture modes in polycrystalline aggregates. The fracture behaviour of single crystals subjected to tension depends on their crystalline structure, their orientation with respect to the tensile axis and the testing temperature. The types of fracture generally encountered are: (i) brittle cleavage with little or no plastic deformation; (ii) substantial necking culminating in rupture (100% reduction in area); (iii) necking with ductile crack initiation at a late stage; (iv) localized shearing along one or more coarse slip bands, ending in crack propagation either along one band or across the bands. Pure face-centred cubic crystals normally fracture either by rupture or by extensive necking prior to crack

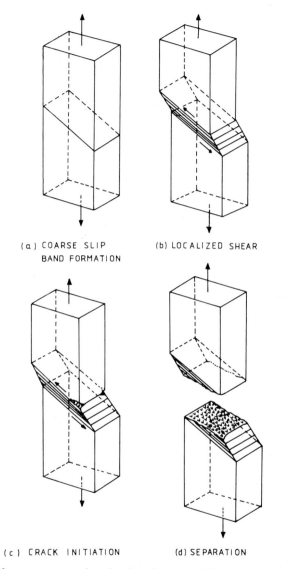

(a) COARSE SLIP
BAND FORMATION

(b) LOCALIZED SHEAR

(c) CRACK INITIATION

(d) SEPARATION

Fig. 4.4 The fracture process in a hardened crystal. (After R. J. Price and A. Kelly, *Acta Metall.* **12**, 979–997 (1964).)

initiation. For a given metal, crack development in the neck becomes more predominant as the testing temperature is lowered. From tests on aluminium and copper crystals described by Beevers and Honeycombe (1959) it was shown that two or more slip systems operate in the neck, but when the neck is developed coarse slip occurs on one system. In copper crystals necking and localized shearing also occurs, but the necking is less extensive than in aluminium. At a later stage in the process a crack develops in the centre of the necked region, which, with further straining, leads to separation. For certain age-hardenable aluminium-alloy single crystals hardened so that Guinier Preston I (GP I) or GP II zones are present, little or no necking occurs prior to fracture. The fracture process consists of three stages: coarse slip-band formation, shear along one band and crack propagation across the band (Beevers and Honeycombe, 1961). The coarse slip bands form abruptly and there is a drop in load. It has been observed that the bands form at a constant resolved shear stress. After the formation of these bands, deformation is confined to localized shearing along one band, which also occurs under a decreasing load. The crack is initiated at the edge of this band. The progressive nature of this type of fracture is shown schematically in Fig. 4.4.

Body-centred cubic crystals generally either cleave or neck down to a chisel edge (100% reduction in area). Much of the work on body-centred cubic crystals has been concerned with iron. It was shown by Allen et al. (1956) that high-purity iron crystals fracture in a brittle manner below a temperature of $-196\,^\circ$C, whereas above this temperature fracture occurs by rupturing for all orientations. Close-packed hexagonal metals either fracture by cleavage or by necking, depending on the metal and the testing temperature. An example of the temperature dependence of fracture mode is shown by zinc. At a temperature of $-196\,^\circ$C zinc cleaves, whereas at a temperature of $300\,^\circ$C rupture occurs. Cadmium and tin also fracture by rupture at room temperature. An interesting and complex necking behaviour is sometimes observed with tin single crystals. Initially a neck is formed by slip on two systems simultaneously. Then a further elongation occurs in the necked region, before a second neck forms and the crystal separates by rupture. Similar behaviour to this was reported for zinc by Schmid and Boas (1950).

4.2   Fracture in polycrystalline metals

Detailed work by Puttick (1959) and Rogers (1960) on polycrystalline copper tensile specimens established beyond any doubt that void nucleation and growth is the most common mechanism for ductile fracture.

In tough-pitch copper Puttick observed that voids nucleated at oxide inclusions either by decohesion at the particle matrix interface or by inclusion cracking. Rogers carried out his experiments on oxygen-free high-conductivity copper. In specimens heated in hydrogen (to weaken grain boundaries) voids were observed at grain-boundary triple points and grain boundaries. As necking progressed, larger voids formed in the area of minimum section. Intense shear bands formed at large strains. For specimens that were heated in a vacuum, void nucleation occurred within the grains. There is evidence that in other metals, particularly at higher temperatures, void nucleation within the grains can occur.

Broek (1971, 1973) showed that voids may form at large particles, even at small strains in aluminium alloys, and microcracks formed by particle cracking for long slender cracks. Further confirmatory work on aluminium alloys was carried out by Hahn and Rosenfield (1975).

4.2.1 Fracture in steels

In carbon steels fractures either develop from microcracks associated with inclusions (see Wilson, 1971) or from cracking in pearlite colonies. Ductile manganese sulphide inclusions are also often the origin of microvoids in steels. Such inclusions may already be cracked by prior mechanical working processes. In the development of modern steels (e.g. high-strength low-alloy (HSLA) steels) inclusion shape control is of great importance, as described by Pickering (1978). The usual problem with HSLA steels is lack of through-thickness ductility and low impact toughness in the through-thickness direction caused by inclusions that have elongated in the rolling or working direction. Honeycombe (1981) lists the five inclusion types according to deformability. This categorization is based on one used by Kiessling and Lange (1966).

(1) Alumina and calcium aluminates, which are brittle and are present because of deoxidation practice.

(2) Spinel-type oxides, which deform only at temperatures above about 1200 °C.

(3) Silicates of calcium, managanese, iron and aluminium. The deformability of this inclusion type increase with increasing temperature.

(4) FeO and MnO, which are plastic at room temperature but become brittle above about 400 °C.

(5) MnS, whose deformability increases with decreasing temperature. The morphologies of MnS inclusions are normally divided into three types (see e.g. Honeycombe, 1981; Leslie, 1982). Type I inclusions are globular and form in rimming steels; those of type II are formed in

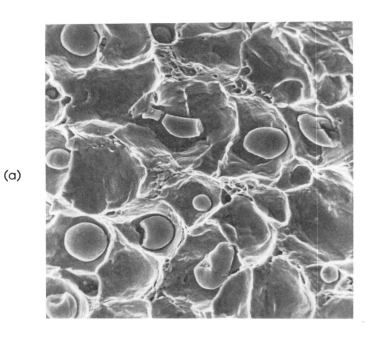

(a)

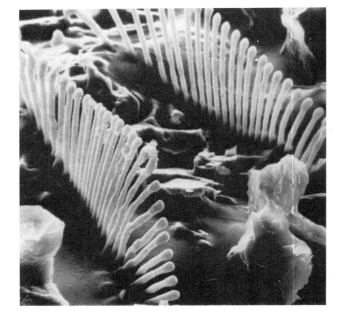

(b)

(c)

Fig. 4.5 Types of manganese sulphide inclusion in steels: (a) type I × 415; (b) type II × 825; (c) type III × 2000. (Courtesy T. J. Baker, see also T. J. Baker and J. A. Charles, *J. Iron Steel Inst.* **210**, 702–706 (1972).)

killed steels and have an interdendritic structure, while type III inclusions are random angular particles that are formed in fully deoxidized steels. These three types of MnS inclusion are shown in Fig. 4.5.

When inclusions form with the "type II MnS" morphology there is a very marked difference between the ductility and toughness in the rolling and through-thickness directions. Methods for overcoming this lack of through-thickness ductility, caused by sulphides and oxides, are by adding calcium, zirconium and cerium. These additions apparently decrease the plasticity of the inclusions forming globular shapes.

Porter *et al.* (1978) studied the fracture behaviour of pearlites ranging in interlamellar spacing from a relatively coarse 400 nm to a very fine spacing of 90 nm. These researchers carried out *in situ* tensile deformation in a high-resolution scanning electron microscope. In all cases localized shear in the pearlite colonies was observed. Cracks appeared to be either associated with the bands of localized shear in the pearlite or the inclusions.

In coarse pearlites deformation appeared to be homogeneous; however, at higher strains deformation becomes inhomogeneous, which manifests itself as intense shear bands intersecting the previously continuous lamellae aligned along the tensile axis. In many cases these shear bands pass through growth faults, this is indicated by the letter A in Figs. 4.6(a), (c) and (d). Figure 4.6(b) illustrates the nature of cementite deformation. It can be seen that the lamella at A is protruding from the ferrite. Final fracture occurs along the bands of intense shear.

Figure 4.7 illustrates a suggested mechanism for shear cracking in pearlite. Miller and Smith (1970) consider that for shear cracking to occur in pearlite it is necessary to have a tensile stress acting along the cementite plates. After one cementite plate has cracked, plastic shear deformation occurs in adjacent lamellae until further cementite plates crack. At about this stage it is possible for one part of the pearlite colony to move bodily with respect to the other as shown in Fig. 4.7. The microvoids grow in size with increasing deformation until a thin shear fracture is obtained. The dimple patterns produced by these fractures are very shallow, indicating that the band of intense shear deformation is narrow.

The cracking of fine pearlites is normally much more complex than that of coarse pearlites. Unlike coarse pearlite, originally continuous lamellae of cementite can neck down and break into numerous fragments, as shown in Fig. 4.8(a) at point A. Although the cementite lamella has necked down, it is interesting to note that the ferrite/cementite interface still appears to be intact. Microvoids can be seen at point B in Fig. 4.8(b).

Figure 4.9 shows a detail of a shear band in fine pearlite. The width of the shear band is large with respect to the interlamellar spacing, unlike the case of coarse pearlites. Kinking, which is sometimes observed in pearlite subjected to compression, has not been found in the tensile tests described above.

As ferrite is more ductile than pearlite, it should be expected that cracks will originate at the pearlite colonies; this is indeed the case. The proeutectoid ferrite can withstand extensive plastic shear strains, and cracks always begin at either pearlite colonies or inclusions.

4.2.2 Fracture in nickel-based superalloys

Kikuchi et al. (1981) have shown that the size, spacing and morphology of grain-boundary carbides, primarily of the type $M_{23}C_6$, influenced the nucleation of voids during cold working of Astroloy (primarily a nickel–cobalt–chromium–molybdenum alloy). These workers found that the cavity spacing was proportional to the carbide spacing. Further, it was shown that voids form at the ends of carbide particles. Voids nucleated primarily at the

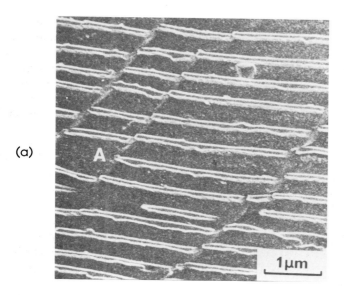

(a)

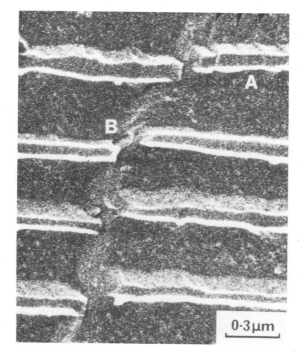

(b)

(c)

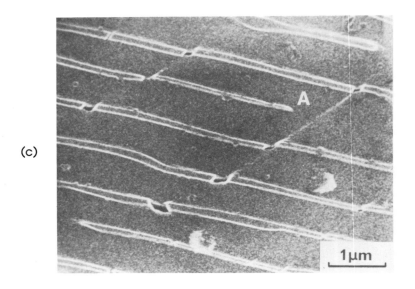

(d)

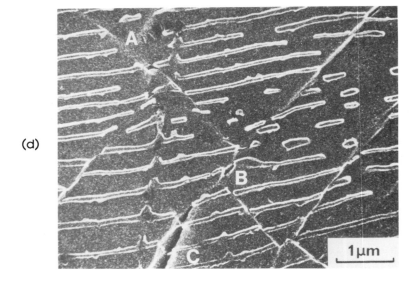

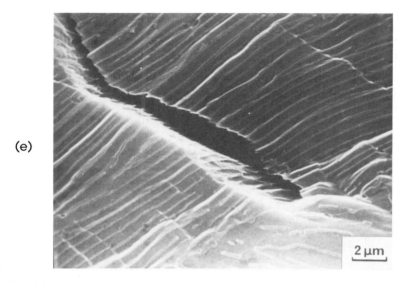

(e)

Fig. 4.6 Intense shear bands in coarse pearlite: (a) shear bands with the lamellae nearly parallel to the tensile axis; (b) a higher resolution view of part of a shear band, (c) cavity formation; (d) initiation of cracking; (e) serrated shear crack. (Courtesy G. D. W. Smith, from D. A. Porter *et al.*, *Acta Metall.* **26**, 1405–1422 (1978).)

intersection of slip bands with grain-boundary carbides and numerous slip bands intersected the grain boundaries near the particles.

It was also observed that voids form primarily on the boundaries oriented parallel to the tensile axis. A series of typical voids are shown in Fig. 4.10.

4.3   Void-nucleation mechanisms

A number of mechanisms for void nucleation have been proposed, these fall broadly into two categories: (*a*) homogeneous and (*b*) heterogeneous nucleation. Concern is with dislocation interactions and vacancy migration for homogeneous nucleation mechanisms, whereas heterogeneous nucleation mechanisms involve inclusions, second-phase particles or some other structural discontinuities.

4.3.1   Dislocation models for homogeneous nucleation

It has been proposed that voids could form by the condensation of vacant lattice sites. However, except at elevated temperatures, where vacancy diffusion is appreciable, voids cannot form by this mechanism.

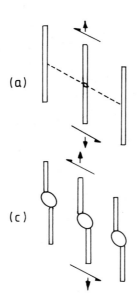

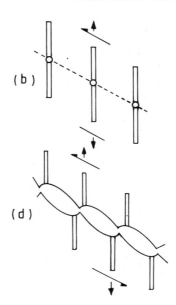

Fig. 4.7 Suggested mechanism for shear cracking of pearlite: (a) cracking of one cementite plate; (b) shear zone forms in the ferrite when adjacent plates are fractured; (c) void formation; (d) void coalescence. (After L. E. Miller and G. C. Smith, *J. Iron Steel Inst.* **208**, 998–1005 (1970).)

An alternative means of homogeneous nucleation is by the coalescence or interaction of dislocations. Mechanisms suggested by McLean (1962) for dislocation coalescence are shown in Fig. 4.11. In Fig. 4.11(a) a cluster of dislocations is piled-up against a boundary to give a crack. Figure 4.11(b) shows how two edge dislocations of opposite sign attract each other, forming a microcrack. The third mechanism of void nucleation is shown in Fig. 4.11(c); here two dislocations on different planes intersect. These methods of void nucleation have been observed in certain cases.

4.3.2 Dislocation models for void nucleation at particles

A number of dislocation models have been suggested by which it is possible to nucleate voids at particles. Ashby (1966) assumed that prismatic dislocation loops would be formed at the particle–matrix interface where the tensile stresses are high. These vacancy loops will glide away from the interface and new loops will form.

For aluminium alloys, Broek (1971) modified this model to allow for cross-slip. Cross-slip is followed by primary slip of the screw dislocations, which

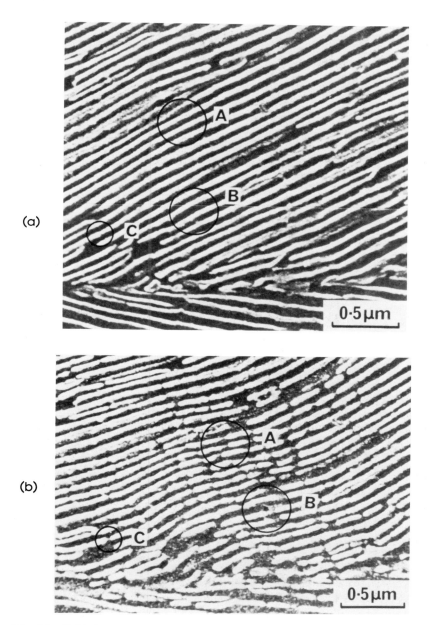

Fig. 4.8 Cold-drawn material before (a) and after (b) deformation, clearly showing cementite plate fragmentation. (Courtesy G. D. W. Smith, from D. A. Porter *et al.*, *Acta Metall.* **26**, 1405–1422 (1978).)

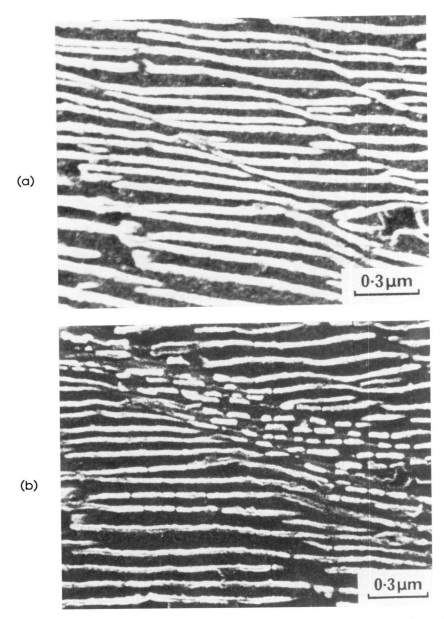

(a)

(b)

0·3 μm

0·3 μm

Fig. 4.9 Cold-drawn material showing a wide shear band formed by deformation. (Courtesy G. D. W. Smith, from D. A. Porter *et al.*, *Acta Metall.* **26**, 1405–1422 (1978).)

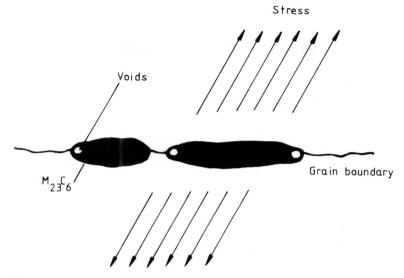

Fig. 4.10 Void nucleation at the ends of carbide particles along grain boundaries. (After M. Kikuchi *et al.*, *Acta Metall.* **29**, 1747–1758 (1981).)

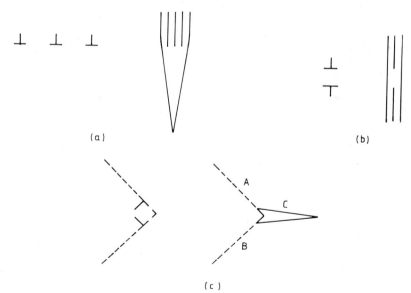

Fig. 4.11 Modes of homogeneous void nucleation by dislocation interaction. (After D. McLean, *Mechanical Properties of Metals.* Copyright 1962 John Wiley and Sons Ltd.)

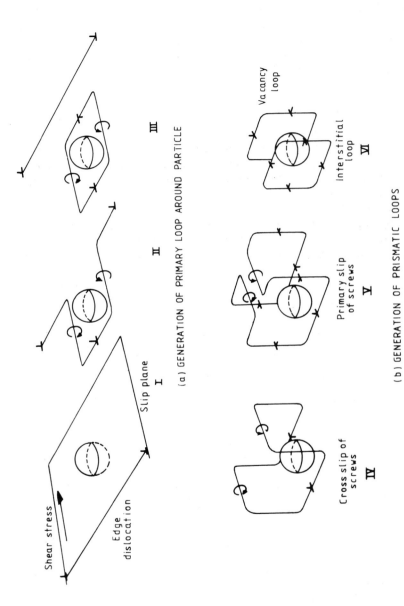

Shear stress

Slip plane

Edge dislocation

I

II

III

(a) GENERATION OF PRIMARY LOOP AROUND PARTICLE

Cross slip of screws

IV

Primary slip of screws

V

Vacancy loop

Interstitial loop

VI

(b) GENERATION OF PRISMATIC LOOPS

Fig. 4.12 Dislocation model for void nucleation at particles involving cross slip and annihilation of screw dislocations. (After D. Broek, A study on ductile fracture. NRL-TR-71021-U (1971).)

leads to their eventual annihilation leaving two prismatic loops, as shown in Fig. 4.12. The vacancy loop initiates the void, and clearly a number of such loops are required for a void to be nucleated. In practice, dislocation tangles form around particles and the local strain fields will be extremely complex because of this. Nevertheless, in these cases dislocation pile-ups will obviously occur and microcracks will form.

4.4 Void coalescence and fracture

The two types of ductile fracture that are most often observed in rod tensile specimens are the cup-and-cone and the double-cup fracture. Referring to Fig. 4.13, it is assumed that a lenticular crack has formed by void coalescence. Further growth of the crack can take place by general deformation or localized shear deformation. Localized shear is favoured by reduced strain-hardening capacity and plane-strain conditions, and thus it should occur later in the process. In a pure metal, growth is expected to be like this resulting in a double-cup fracture. In an impure or commercial metal growth

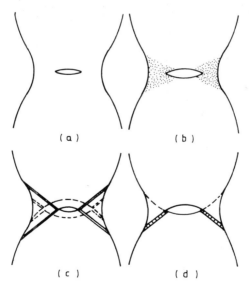

Fig. 4.13   Crack growth in tensile fracture: (a) lenticular crack; (b) growth by general plastic deformation; (c) growth by localized shear: (d) cup-and-cone fracture. (After G. Y. Chin *et al.* Reprinted with permission from TRANSACTIONS OF THE METALLURGICAL SOCIETY, Vol. 230, pp. 1043–1048, (1964), a publication of The Metallurgical Society, Warrendale, Pennsylvania.)

occurs by localized shear and new voids are generated. As deformation continues, voids are nucleated along these bands. The final failure is along just such a weakened band.

Zener (1948) suggested that final separation in a tensile test-piece occurs by adiabatic shear, leading to the characteristic conical shape of a cup-and-cone fracture. However, adiabatic shear has been discounted by some workers.

From all the metallographic evidence, the sequence of events that leads to ductile fracture in a tensile test-piece can be summarized as follows (Oyane, 1972):

(i) voids form after a critical amount of strain;
(ii) at the maximum in the applied load necking begins with further straining;
(iii) at a critical volume fraction of voids or at a critical mean free path between the voids, strain concentrates in narrow bands connecting the voids;
(iv) separation occurs along these bands.

It is appealing to consider void coalescence as a gradual process involving the thinning down of the ligaments between the voids with increasing strain. Final separation would occur when the ligaments have reduced to zero width. This mode is shown schematically in Fig. 4.14. McClintock (1968) showed in

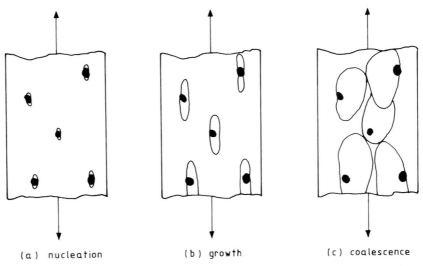

(a) nucleation   (b) growth   (c) coalescence

Fig. 4.14 Void growth from particles and eventual coalescence by the continued necking of the ligaments between the voids.

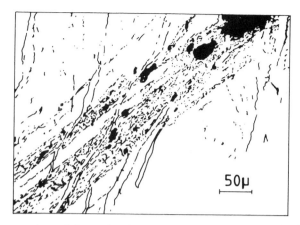

Fig. 4.15 Intense shear deformation in oxygen-free high-conductivity copper. (After H. C. Rogers. Reprinted with permission from TRANSACTIONS OF THE METALLURGICAL SOCIETY, Vol. 218, pp. 498–506 (1960), a publication of The Metallurgical Society, Warrendale, Pennsylvania.)

detailed calculations using his theory that fracture strains an order of magnitude greater than those observed experimentally are obtained when void coalescence by this internal-necking mechanism is assumed. Another approach to void coalescence, made by Schmitt and Jalinier (1982), is that fracture will occur when the volume fraction of voids reaches a critical value that is a characteristic of the material. However, it is unclear what effect the individual sizes of the voids would have. Figure 4.15 shows an intense shear band that typically forms between voids in high-conductivity copper. From this evidence, it would seem that separation results after intense shear between voids. The role of intense shear deformation between free surfaces will be enlarged upon in later chapters.

4.5 Fractography

Valuable information can be obtained about fracture surfaces formed by ductile failure by studying them in an electron microscope. Void initiation sites at particles can often be observed if care is taken. Void nucleation and growth produce very characteristic fractographic details. The fracture surface consists of what are referred to as dimples. The dimples themselves are very shallow holes. Although not all particles will produce voids, the fracture will follow a path through the particles which have produced voids. This

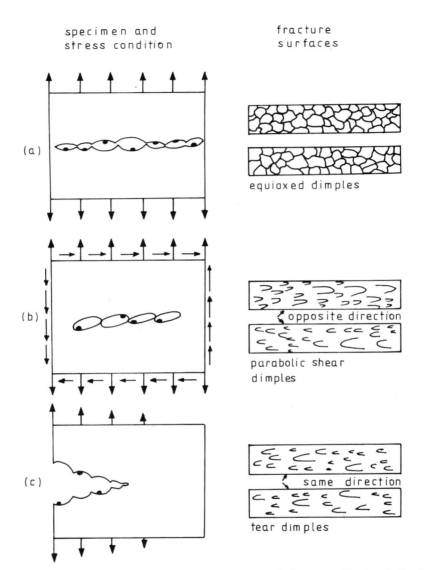

specimen and
stress condition

fracture
surfaces

(a)

equiaxed dimples

(b)

opposite direction

parabolic shear
dimples

(c)

same direction

tear dimples

Fig. 4.16 Categories of dimples according to the applied stresses. Equiaxed dimple patterns occur when the applied stresses are tensile, and parabolic dimples are formed when the applied stress is a shear stress. So-called tear dimples result from a non-uniform applied tension.

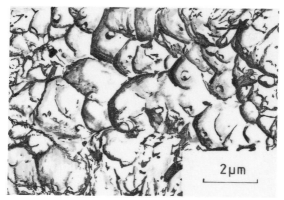

Fig. 4.17 A dimple pattern showing that the dimples are initiated at particles in an aluminium alloy. (After D. Broek, *Elementary Engineering Fracture Mechanics.* Martinus Nijhoff, The Hague, 1982.)

Fig. 4.18 A sketch based on a negative replica of a parabolic dimple pattern in oxygen-free high-conductivity copper. (After H. C. Rogers. Reprinted with permission from TRANSACTIONS OF THE METALLURGICAL SOCIETY, Vol. 218, pp. 498–506 (1960), a publication of The Metallurgical Society, Warrendale, Pennsylvania.)

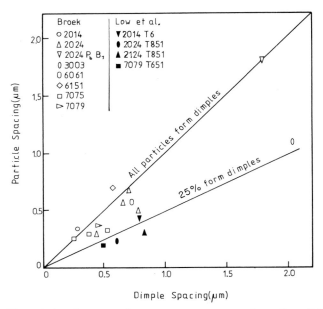

Fig. 4.19 The relation between the centre-to-centre spacing of particles and the dimple spacing on the fracture surface. Results are taken from Broek (1973) and Low *et al.* (1972), combined, the plot is from Hahn and Rosenfield (1975).

behaviour leads to the characteristic irregular fracture surface, as described by Broek (1982).

Three categories of dimple pattern are observed; these are equiaxed, parabolic and tear dimples. Equiaxed dimples are formed in predominantly tensile stress fields, whereas parabolic dimples are formed by shear deformation. Figure 4.16 shows how the different categories of dimples can form.

Figure 4.17 shows a dimple pattern in an aluminium alloy. Some of the particles from which the voids initiated can be seen. Figure 4.18 is an example of a negative replica of parabolic dimple patterns on the fracture surface of high-conductivity copper. Parabolic dimple patterns can be observed on the cup walls in tensile cup-and-cone failure.

The relationships between the centre-to-centre spacing of particles and dimple spacing should be linear. Indeed this is the case, as shown in Fig. 4.19. Although there is a great deal of scatter, it is interesting to note that a one-to-one relation obtains when all particles form dimples, but when only a proportion of the particles nucleate dimples the relation is less steep.

Theories of Ductile Fracture
I: Void Dynamics

5.1 Void nucleation

5.1.1 Theories of void nucleation

As we have seen, there is evidence for the existence of microcracks in some metals, often caused by prior processing, and these usually originate by decohesion of particle–matrix interfaces or by particle cracking. If there are no microcracks present before plastic deformation then nucleation can occur after relatively small strains.

Theories of void nucleation can be divided into two broad categories: (a) theories that treat the problem as one of linear elasticity; and (b) those in which voids are assumed to be nucleated by plastic deformation (caused by dislocation glide and interaction).

In the last chapter we briefly observed that Goodier (1933) calculated the stress states in the vicinity of a particle embedded in a matrix otherwise subjected to a uniform stress. Gurland and Plateau (1963) extended this approach and proposed a criterion for void nucleation at a spherical particle in which the stiffness of the particle is greater than that of the matrix. Under these circumstances, when the matrix is subjected to a tensile load, tensile stresses are produced at the poles of the particle. Conversely, for a particle with a stiffness less than that of the matrix, stresses build up at the equator of the particle. Gurland and Plateau's approach is very similar to the classical Griffith (1921) approach to brittle fracture, in that a stress is derived from an energy balance in which it is assumed that the strain-energy relief due to void nucleation, $a^3(\alpha\sigma)^2/E$ must at least be equal to the surface energy required to create the new surfaces, γa^2. From this balance it is found that the matrix stress for void initiation is

$$\sigma = \frac{1}{\alpha}\left(\frac{E\gamma}{a}\right)^{1/2}, \tag{5.1}$$

where a is the diameter of the particle, γ is the surface energy, E is Young's modulus and α is a stress concentration factor.

In the case of elongated particles, the Gurland–Plateau energy balance may still be used when account is taken of the length and diameter of the particle. For this case, the matrix stress required for void nucleation is derived as follows:

$$\frac{(\alpha\sigma)^2}{E}\, a^2 \lambda = \gamma a^2,$$

from which

$$\sigma = \frac{1}{\alpha}\left(\frac{E\gamma}{\lambda}\right)^{1/2}. \tag{5.2}$$

It is interesting to observe that the stress appears to be only dependent on the particle length and not on its diameter. Rosenfield (1968) suggested that the energy term γ consists of three components: the surface energy of the matrix γ_m, the surface energy of the particle γ_p, and the interface energy γ_{mp}. Then for interface decohesion

$$\gamma = \gamma_m + \gamma_p - \gamma_{mp}, \tag{5.3}$$

and for particle fracture $\gamma = 2\gamma_p$.

Despite the detailed work that has gone into the above elasticity analyses, it is clear that at best they can only be approximations, as we are normally concerned with large plastic deformations of the matrix. The experimental evidence of Palmer et al. (1966) shows that voids form in alloys along bands of intense shear at stresses of about $\frac{1}{20}$ of that given by (5.1). These researchers examined dispersions of silica, beryllia and alumina in copper produced by internal oxidation. This work shows conclusively that plastic strain is required before void nucleation can occur.

As an indication of just how different precipitates and particles can be in a metal matrix, we shall look briefly at precipitation-hardenable alloys and the classes of precipitate that are possible (see Martin, 1980).

In these alloys the interface between a precipitate and the matrix may be fully coherent, semicoherent or completely incoherent. A coherent precipitate is one in which there is atomic matching across the interface between the particle and the matrix (Fig. 5.1a). Any misfit along the interface is accommodated by elastic coherency strains. A semicoherent precipitate is

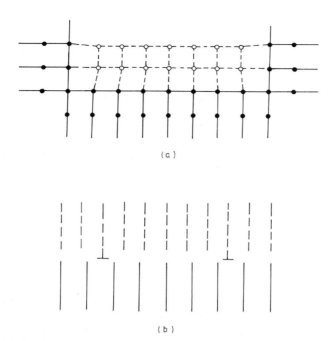

(a)

(b)

Fig. 5.1 (a) A coherent interface between two crystals with slightly different atomic spacings, the strains set up at the interface are accommodated by elastic coherency strains. (b) A quasicoherent interface in which there is a significant mismatch in atomic spacings, this mismatch is accommodated by a regular array of dislocations. (After J. W. Martin, *Precipitation Hardening*. Copyright 1968 Pergamon Books Ltd.)

one in which the regions of misfit are accommodated by an array of dislocations at the interface, as shown in Fig. 5.1(b). Finally, the interface can be completely incoherent, in which case there is no matching across the precipitate–matrix interface. In general, it appears that dislocations can cut through coherent and semicoherent precipitates, but this is usually not so for incoherent particles. Thus any method for the dispersion of fine insoluble particles in a matrix, by for example powder metallurgy, will increase the strength of the metal. To illustrate the effects of particle size and distribution on the ease of movement of dislocations and hence strength we will, briefly, describe alloys that may be precipitation hardened. Figure 5.2(a) shows a dislocation that has taken up its minimum-energy configuration, with a large average radius R, surrounded by individual solute atoms or small clusters of solute atoms with a mean spacing Λ, such that $\Lambda \ll R$. In this case the individual atom clusters have very weak stress fields. As shown in Fig. 5.2(b),

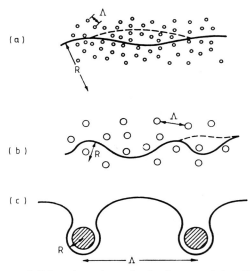

Fig. 5.2 Movement of dislocations through a lattice containing dispersed particles of differing sizes: (a) the particles are very small and are either very small zones or individual atoms with $\Lambda < R$; (b) intermediate sized particles where $\Lambda \sim R$; (c) large precipitates with $\Lambda \gg R$. (After D. Hull, *Introduction to Dislocations*, 2nd edn. Copyright 1975 Pergamon Books Ltd.)

with intermediate-sized particles, where $\Lambda \approx R$ (for example with semicoherent precipitates), the dislocations can cut the particles, as is the case for smaller particles, but a greater applied stress is required because of the greater elastic stress fields of the particles. It is worthy of note here that an alloy containing a semicoherent precipitate is at its peak strength. An overaged alloy is typified by large, widely separated, precipitated particles with $\Lambda \gg R$. In this last case, shown in Fig. 5.2(c), a dislocation passing in the vicinity of an array of precipitates can overcome its line tension and bow around the particles. As each dislocation bows around the particles, it forms a series of loops as described by Orowan (1947) until plastic relaxation processes begin to operate. In the practical case of a particle embedded in a plastic flowing matrix, very complex dislocation tangles can result, particularly with the possibility of particle cutting by dislocations. To complicate matters further, particles can also produce dislocations by punching out prismatic dislocation loops to relieve the internal stresses. A good example of this is shown in Fig. 5.3, which is a transmission electron micrograph of prismatic loops punched out at a carbide precipitate in iron by cooling. Ashby (1966) developed a dislocation model for void nucleation that illustrated the importance of the

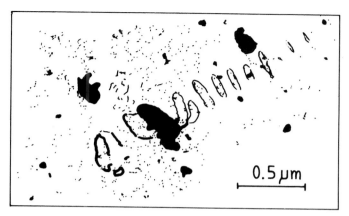

Fig. 5.3 TEM of prismatic dislocation loops punched out at a carbide precipitate. (Drawing based on an original by D. Hull, *Introduction to Dislocations*, 2nd edn. Copyright 1975 Pergamon Books Ltd.)

relief of tensile stresses at the poles of a particle in a shear band. If a spherical hole is present in the shear band then it will become elliptical. Now if the hole contains a particle that is well-bonded to the matrix then, as the band deforms, localized stresses will be generated at the particle (see Fig. 5.4a). As plastic deformation continues, the internal stresses in the vicinity of the particle exceed the yield stress of the matrix, and prismatic vacancy loops are punched out in the tensile-stressed regions and interstitial loops are produced in the matrix in the areas that are subjected to compressive stresses. The vacancy loops will eventually coalesce at the interface, forming a microvoid. Ashby showed that the matrix strain required for microvoid nucleation is

$$\varepsilon = \frac{CbL\sigma_{cr}}{a},\qquad(5.4)$$

where C is a constant, b is Burgers vector, L is the pile-up size, a is the particle size and σ_{cr} is the critical stress required for nucleation.

Because prismatic loops tend to be sessile, modifications to Ashby's model based on suggestions by Hirsch (1957) have been made by, for example, Gleiter (1967). For elongated particles it is assumed that cleavage fracture of the particle occurs to relieve the stresses produced by dislocation pile-ups. Figure 5.5 shows a particle subjected to such pile-up stresses. If we assume that there are n dislocations in the pile-up of length L then the shear stress of the pile-up is $\tau = n\tau_a$, where τ_a is the applied stress. The number of

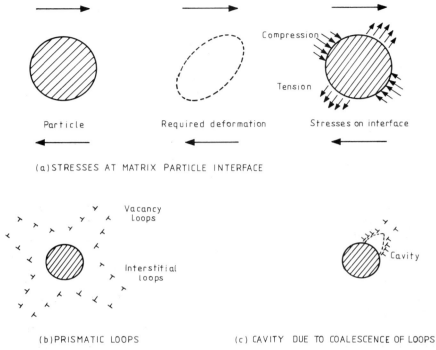

(a)STRESSES AT MATRIX PARTICLE INTERFACE

(b)PRISMATIC LOOPS   (c) CAVITY DUE TO COALESCENCE OF LOOPS

Fig. 5.4 Void nucleation after Ashby (1966). (From D. Broek, A study on ductile fracture. NRL-TR-71021-U (1971).)

dislocations n is given approximately by

$$n \sim \frac{2L\tau_a}{Gb},$$ (5.5)

where G is the shear modulus and b is Burgers vector. If the inclusion cracks then the shear stress, τ, exerted on the particle exceeds the critical value for particle fracture τ_{cr}. Therefore a condition for void initiation is

$$\tau_a = \left(\frac{Gb\tau_{cr}}{2L}\right)^{1/2}.$$ (5.6)

A number of researchers have studied microvoid nucleation by particle cracking, for example, Kelly and Davies (1965), Gell and Worthington (1966) and Barnby (1967). In particular, Barnby derived an equation for the critical fracture stress of carbide particles in stainless steel.

It is difficult to state with any certainty whether a characteristic nucleation

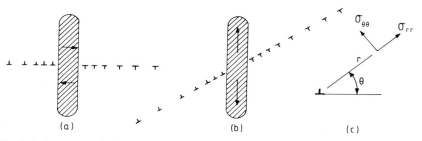

Fig. 5.5 Fracture of elongated particles due to dislocation pile-ups. (After D. Broek, A study on ductile fracture. NRL-TR-71021-U (1971).)

strain must be achieved before void nucleation can occur. From the above theories, it would seem that a critical strain is required. But, practically, we must remember the stochastic distribution of particle size and so forth. It follows that there will always be some uncertainty attached to the evaluation of a nucleation strain. Goods and Brown (1979) in their literature review favour the concept of a void-nucleation strain, and if one is concerned with the nucleation of a single void then it seems sensible that it should nucleate at a critical strain.

Abe *et al.* (1983) have applied the rigid-plastic finite-element method to a matrix containing a regular array of inclusions. Both the inclusions and the matrix are assumed to be rigid plastic. These researchers also investigated the effects of inclusion aspect ratio. Unlike the elastic strains in elastic inclusions, Abe *et al.* found that the plastic strain within rigid-plastic inclusions is not uniform. Large plastic strains are possible both within an inclusion and at the interface of the inclusion with the matrix, depending on the respective mechanical properties. Clearly, the results of this type of work will give more accurate predictions of the conditions for the nucleation of micro-voids by plastic deformation in the vicinity of inclusions.

5.1.2 Superimposed hydrostatic pressure and void nucleation

Brown and Stobbs (1976) developed a model for void nucleation in a matrix that contains hardening particles (silica particles in a copper matrix). In the analysis the contributions to hardening are derived in terms of a critical strain. When the applied strain is the same as the critical strain, void nucleation occurs under the action of a superimposed hydrostatic pressure when

$$\sigma_f + \sigma_c + p \geqslant \sigma_1, \tag{5.7}$$

where σ_f reflects the various contributions of the microstructure, such as particle volume fraction and dislocation density, σ_c is the local stress, p is the hydrostatic stress and σ_1 is the particle–matrix interfacial strength. Goods and Brown (1979) found for ferrous materials that the critical strain for void nucleation is given by

$$\varepsilon^{1/2} \geqslant K(\sigma_1 - p), \qquad (5.8)$$

where K is a constant for a given volume fraction of particles and particle size.

Figure 5.6 shows $\varepsilon^{1/2}$ versus the hydrostatic stress for spheroidized steel from the results of French and Weinrich (1975) and Argon and Im (1975). If, following Goods and Brown, we take p to be 2000 MPa then we obtain a tolerably linear relation, bearing in mind the assumptions made in the analysis. We can see quite clearly that as the hydrostatic stress (tension) increases, the nucleation strain decreases.

Although complex, the model of Brown and Stobbs does give the expected trend of the effects of hydrostatic pressure on nucleation strain. Obviously at the one extreme of very high hydrostatic tension voids nucleate immediately plastic deformation begins, and at the other extreme of very high pressures void nucleation is suppressed completely.

The above description of theories of void nucleation emphasizes the complexity that confronts researchers, as mentioned by Wilsdorf (1983). The problem is to marry microstructural details such as properties and size of

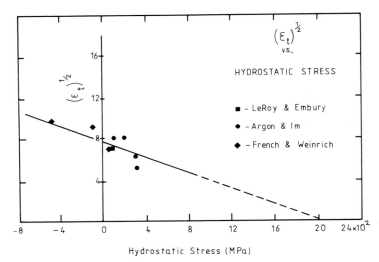

Hydrostatic Stress (MPa)

Fig. 5.6 Nucleation strain versus hydrostatic pressure. (After S. H. Goods and L. M. Brown, *Acta Metall.* **15**, 1213–1228 (1979); data points are from French and Weinrich (1975) and Argon and Im (1975).)

inclusions with the macroscopic applied stress. The true complexity of the process of void nucleation is shown in Table 5.1. Here we can see some of the many microstructural and macroscopic variables involved in ductile fracture involving second-phase particles or inclusions.

Table 5.1

Variables in ductile fracture mechanisms involving second-phase particles (after H. G. F. Wilsdorf, *Mat. Sci. Engng* **59**, 1–39 (1983))

Shape of specimen	Cylindrical
	Sheet
	Other
Particles	Volume fraction
Type of particle	Inclusion
	Precipitate
	Dispersion
Particle shape	Spherical
	Elongated
Particle size	<10 nm; 0.05–1 μm; >1 μm
Particle location	Matrix
	Grain boundary
	Spacing between particles
	Orientation
Grain structure of matrix	Size
	Shape
	Preferred orientation
	Grain boundaries
Free surface energy	Matrix
	Particle
	Matrix–particle
State of stress	Uniaxial
	Triaxial
	Hydrostatic
	Normal
Strain	Magnitude
	Rate
Stress	Yield stress
	Flow stress
	Fracture stress
Strain hardening	Dislocation cell structure
	Deformation mode due to stacking fault energy

After Wilsdorf (1983).

5.2 Theories of void growth

5.2.1 Two-dimensional theories

With plastic deformation a regular array of holes of initially circular cross-section will, in general, become ellipsoidal in cross-section. Although the solution for a circular hole deforming in a plastic matrix is known, the corresponding solution for holes with ellipsoidal cross-section is not. McClintock (1968) approximated Berg's (1962) solution for deformation of ellipsoidal holes in a viscous material to a plastic strain-hardening material. For a plastically deforming matrix containing a regular array of cylindrical holes of elliptical cross-section, McClintock derived approximate expressions for the rate of damage assuming that there is no rotation of principal axes. Each hole is assumed to be within an identical cell as shown in Fig. 5.7. It is further assumed that the matrix deforms in plane strain and that the holes are still small enough that their interaction may be neglected. Following McClintock's notation, we take the dimensions of a cell to be l_a and l_b, which correspond to the respective hole spacings, and take a and b to the corresponding semi-axes of the holes. Two hole-growth factors can be defined.

$$\left. \begin{aligned} F_a &= \frac{a/l_a}{a_0/l_a^0}, \\[2mm] F_b &= \frac{b/l_b}{b_0/l_b^0}, \end{aligned} \right\} \tag{5.9}$$

where a_0, b_0, l_a^0 and l_b^0 are the initial values. Fracture is assumed to occur when there is a complete loss of cross-section. For the two possibilities, the hole growth factors are

$$\left. \begin{aligned} F_a^{\mathrm{f}} &= \frac{l_a^0}{2a_0}, \\[2mm] F_b^{\mathrm{f}} &= \frac{l_b^0}{2b_0}. \end{aligned} \right\} \tag{5.10}$$

Hereinafter we shall assume that the material is a von Mises material that deforms according to the Lévy–von Mises equations:

$$\mathrm{d}\varepsilon_{ij} = \frac{3}{2} \frac{\mathrm{d}\bar{\varepsilon}}{\bar{\sigma}} (\sigma_{ij} - \sigma_{\mathrm{m}}). \tag{5.11}$$

Then, for a circular hole in a non-strain-hardening material subjected to

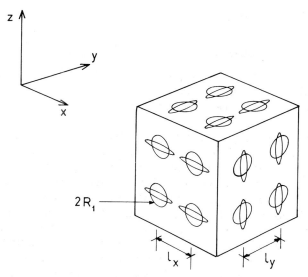

Fig. 5.7 A pattern of cylindrical holes visualized by F. A. McClintock (*J. Appl. Mech.* **90**, 363–371 (1968).)

generalized plane-strain conditions,

$$\ln\left(\frac{R}{R_0}\right) = \sqrt{3}\,\varepsilon_{r\infty}\,\sinh\left\{\frac{\sqrt{3}\,\sigma_{r\infty}}{\sigma_{r\infty} - \sigma_{z\infty}}\right\} + \varepsilon_{r\infty},\qquad(5.12)$$

where R is the hole radius and the strains are logarithmic:

$$d\varepsilon_{\theta R} = \frac{dR}{R} = d\left(\ln\frac{R}{R_0}\right)\qquad(5.13)$$

(see Appendix 5A).

McClintock extrapolated the above equations to give the current mean radius R of elliptical holes in the same material (as shown in Appendix 5A):

$$\ln\left(\frac{R}{R_0}\right) = \frac{\bar{\varepsilon}\sqrt{3}}{2}\,\sinh\left\{\frac{\sqrt{3}}{2}\left(\frac{\sigma_a + \sigma_b}{\bar{\sigma}}\right)\right\} + \frac{\varepsilon_a + \varepsilon_b}{2},\qquad(5.14)$$

where

$$R = \tfrac{1}{2}(a + b).\qquad(5.15)$$

McClintock extrapolated (5.14) to cover this case, as shown in Appendix 5A, assuming power-law hardening

$$\bar{\sigma} = K\bar{\varepsilon}^n;\qquad(5.16)$$

then

$$\ln\left(\frac{R}{R_0}\right) = \frac{\bar{\varepsilon}\sqrt{3}}{2(1-n)} \sinh\left\{\frac{\sqrt{3(1-n)}}{2} \frac{(\sigma_a - \sigma_b)}{\bar{\sigma}}\right\} + \frac{\varepsilon_a + \varepsilon_b}{2}. \quad (5.17)$$

The damage increment for holes with axes parallel to the z-axis coalescing in the b-direction is given by $\mathrm{d}\eta_{zb}/\mathrm{d}\bar{\varepsilon}$ and

$$\eta_{zb} = \ln F_{zb}/\ln F_{zb}^{f}.$$

Also,

$$F_{zb} = \frac{R(1-m)\,e^{-\varepsilon_b}}{R_0(1-m_0)},$$

where $\varepsilon_b = \ln(l_b/l_b^0)$ and the eccentricity $m = (a-b)/(a+b)$.

McClintock found that the major contribution to the growth factor F comes from the radius ratio R/R_0, and the transient effect associated with the shape change, i.e. the eccentricity m, decays rapidly. The damage rate is

$$\frac{\mathrm{d}\eta_{zb}}{\mathrm{d}\bar{\varepsilon}} = \frac{1}{\ln F_{zb}^{f}} \frac{\mathrm{d}}{\mathrm{d}\bar{\varepsilon}} \left\{\ln\left(\frac{R}{R_0}\right) + \ln\left(\frac{1-m}{1-m_0}\right) + \ln e^{-\varepsilon_b}\right\}$$

$$= \frac{\sqrt{3}}{2} \frac{\sinh\left\{\sqrt{3(1-n)}(\sigma_a + \sigma_b)/(2\bar{\sigma})\right\}}{(1-n)\ln F_{zb}^{f}}, \quad (5.20)$$

provided that the second term in $\frac{1}{2}(\varepsilon_a + \varepsilon_b)$ in (5.17) can be dropped; this is in fact the uniform axial strain $-\frac{1}{2}\varepsilon_z$. For the case where the stress ratios remain constant, an approximate closed-form fracture criterion can be integrated from (5.20) at coalescence with $\eta_{zb} = 1$. Then

$$\bar{\varepsilon}_f = \frac{2(1-n)\ln F_{zb}^{f}}{\sqrt{3}\sinh\left\{\sqrt{3}(1-n)(\sigma_a + \sigma_b)/(2\bar{\sigma})\right\}}. \quad (5.21)$$

The criterion given by (5.21) depends on the average of two transverse stresses, which in fact represents the effect of hydrostatic stress on void growth in the two-dimensional void model.

McClintock showed that his theory of coalescence of holes in a plastically deforming mass overestimates substantially the fracture strains observed experimentally in tensile tests, and he therefore concluded that the hole-growth concept can only be considered as an upper limit to the ductility. The same conclusion was drawn by Hancock and Mackenzie (1976) in work carried out on three steels. The differences between theory and experiment may be due to the important fact that coalescence in McClintock's theory is caused by void expansion and, eventually, connection. Some other crucially important processes may occur before the complete loss of cross-section; for

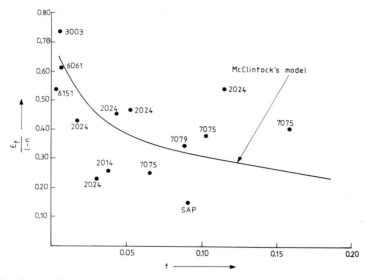

Fig. 5.8   Comparison between fracture strains and volume fraction of particles with McClintock's theory. (After D. Broek, A study on ductile fracture. NRL-TR-71021-U (1971).)

example shear localization between the voids, which is completely ignored in the theory. However, as stated by Guillen-Preckler (1967), although the McClintock theory is approximate, it has the merit of taking into account all the main factors that play a role. Nevertheless, it is dangerous to use this McClintock approach if the voids concerned are not cylindrical.

By simplifying and rearranging (5.21), it can be shown that the fracture strain divided by $(1 - n)$ is a function of the volume fraction f of second-phase particles. Figure 5.8 shows a comparison between the experimental fracture strains for aluminium alloys and McClintock's model. The agreement in most cases is reasonable. It is interesting to note that the theoretical curve follows approximately the same trend as the experimental fracture strains obtained by Edelson and Baldwin (1962) for dispersion hardened copper alloys (Fig. 4.1).

In shear deformation with a superimposed hydrostatic stress there are two possible modes of hole coalescence according to McClintock et al. (1966); these are shown in Fig. 5.9. For hole contact in the longitudinal direction of the shear band the hole-growth factor is

$$\ln F_{\mathrm{L}} = \ln \left(\frac{l_{\mathrm{L}}}{2R_1} \right) = \ln \{(1 + \gamma^2)^{1/2}\} + \ln \left(\frac{R}{R_1} \right), \qquad (5.22)$$

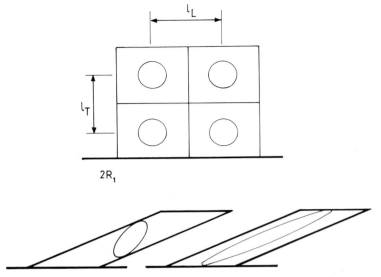

Fig. 5.9 Two modes of hole coalescence in shear with a superimposed hydrostatic pressure according to F. A. McClintock *et al.* (*Int. J. Fract. Mech.* **2**, 614–627 (1966)).

where γ is the shear strain, and for hole contact in the thickness direction of the shear band,

$$\ln F_T = \ln\left(\frac{l_T}{2R_1}\right) = \ln\left(\frac{R}{R_1}\right). \tag{5.23}$$

The radius ratio is given, in a similar way to the previous case, by

$$\ln\left(\frac{R}{R_1}\right) = \frac{\gamma}{2(1-n)}\sinh\left\{\frac{1-n}{\tau}p\right\}. \tag{5.24}$$

A superimposed hydrostatic stress p can influence hole development, as can be seen directly from the above equation. When the normal stress is a hydrostatic tension there is a dramatic decrease in ductility. With moderate amounts of tension, the holes may close, rotate and then reopen, whereas when the hydrostatic stress is compressive the holes remain closed and do not cause fracture. If there is no hydrostatic stress then fracture can only occur in the longitudinal direction of the shear band.

An alternative approach to fracture in plane strain was made by Thomason (1968), who assumed that the material contains a uniform distribution of square prismatic cavities as shown in Fig. 5.10. Fracture occurs when internal necking results from tensile plastic instability in the

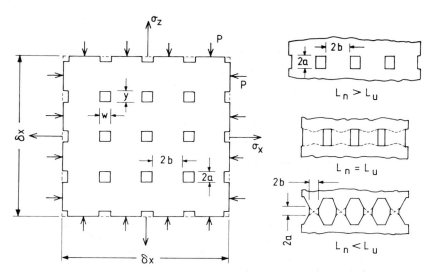

Fig. 5.10 Thomason's model material containing a uniform array of square cavities. (After P. F. Thomason, *J. Inst. Metals* **96**, 360–365 (1968).)

matrix between the cavities, as suggested by Cottrell (1959). Metallurgical results, showing that microvoids, in regions immediately adjacent to ductile fracture surfaces, are still small and widely spaced support Cottrell's concept. A critical-load approach was proposed by Thomason (1968). This procedure provides a necessary condition for instability. A further simplification is that the material is rigid perfectly plastic.

If the distance between voids is large then the mean tensile stress required to cause flow in the neck σ_n may be higher than the yield stress σ_y. The voids will grow until the constraint factor σ_n/σ_y reaches a low-enough value. The element shown in Fig. 5.10 is assumed to be subjected to a tensile stress σ_z and a transverse stress σ_x, which can be tensile or compressive. According to Thomason, a uniform mode of plastic flow will continue until $L_n < L_u$, where $L_n = \sigma_n(\delta x - nw) + mp\,\delta x$ is the load to cause internal necking between a row of cavities and $L_u = \sigma_z\,\delta x$ is the load for uniform flow. Recalling that the conditions for yielding is $\sigma_z = 2k + \sigma_x$ and that the volume fraction of the uniform distribution of cavities is $v_f = (nw/\delta x)^2$, one obtains

$$\frac{\sigma_n}{2k}\left(1 - v_f^{1/2}\right) + \frac{p}{2k} < \frac{\sigma_x}{2k} + 1, \qquad (5.25)$$

where v_f is the volume fraction of voids and k is the shear yield stress. If p is greater than $2k + \sigma_x$ then ductile fracture by void coalescence will not occur.

Thomason (1981) modified his approach and managed to establish a mathematically sound sufficient condition for stability.

This model is very similar to McClintock's in that both assume that fracture occurs when the ligament width between voids reaches some limit. Although this void-growth theory has an obvious appeal, it is a fact that much of the experimental evidence shows that another form of instability occurs in the ligaments. This is a shear instability, as described by Rogers (1960) and Spretnak (1974).

5.2.2 Three-dimensional models

Rice and Tracey (1969) considered the growth of a single spherical cavity in a plastic material subjected to uniaxial tension and a general three-dimensional stress field. Figure 5.11 depicts a void in a matrix which is subjected to a uniaxial tension. Rice and Tracey divided the velocity field into three terms: (1) the uniform strain rate field at infinity, $\dot{\varepsilon}_{ij}^{\infty}$; (2) a symmetrical term related to changes in volume, $\dot{u}_i^D$; and (3) a term related to shape changes occurring without a volume change, $\dot{u}_i^E$. The velocity field then becomes

$$\dot{u}_i = \dot{\varepsilon}_{ij}^{\infty} x_j + D\dot{u}_i^D + E\dot{u}_i^E, \tag{5.26}$$

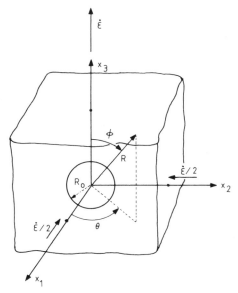

Fig. 5.11 An isolated spherical void within a matrix which is subjected to uniaxial tension. (After J. R. Rice and D. M. Tracey, *J. Mech. Phys. Solids* **17**, 201–217 (1969).)

and incompressibility and symmetry require that $\dot{u}_i^D$ be given by

$$\dot{u}_i^D = \dot{\varepsilon}\left(\frac{R_0}{R}\right)^3 x_i. \tag{5.27}$$

D is taken to be the ratio of the average strain rate of sphere radii to the remote strain rate. The rate of change of shape may be obtained by assuming that spherical surfaces concentric with the void become ellipsoids owing to the field. This rate of shape change can be written in terms of a stream function. Rice and Tracey showed that in tension the volume change is much more important than the shape change of a void, and they also showed that the dilatational amplification factor D increases approximately exponentially with the ratio σ_m^∞/τ, where σ_m^∞ is the remote mean normal stress (see Fig. 5.12). Here the high-triaxiality approximation is given by

$$D = 0.283 \exp\left(\frac{\sigma_m^\infty\sqrt{3}}{2\tau}\right). \tag{5.28}$$

There have been a number of attempts to simplify the Rice–Tracey analysis so that it can be used for a population of voids in a matrix. The stress and strain fields of the voids must obviously be very localized to avoid any interaction.

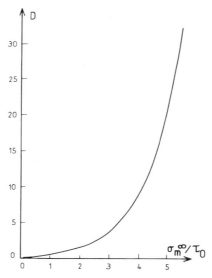

Fig. 5.12 Dilatational amplification factor versus the normalized hydrostatic stress σ_m^∞/τ. (After J. R. Rice and D. M. Tracey, *J. Mech. Phys. Solids* **17**, 201–217 (1969).)

Jalinier (1981) linearized the dilatational amplification factor used by Rice and Tracey to give a relationship between the void growth rate and the applied strain. See also Schmitt and Jalinier (1982).

Hancock and Mackenzie (1976) discussed the Rice–Tracey model and showed that if it is assumed that the failure strain is inversely proportional to the hole growth rate then the failure strain is given by

$$\varepsilon_f = \alpha \exp\left(\frac{-3\sigma_m}{2\bar{\sigma}}\right), \tag{5.29}$$

where α is a constant and $\bar{\sigma}$ is the effective stress. Hancock and Mackenzie make the point that (5.29) is similar to the strain given by a simplified version of McClintock's criterion, i.e. an integrated form of (5.21), which is simplified. A void nucleation strain $\bar{\varepsilon}_n$ can be incorporated in this simplified model as

$$\bar{\varepsilon}_f = \bar{\varepsilon}_n + \alpha \exp\left(\frac{-3\sigma_m}{2\bar{\sigma}}\right). \tag{5.30}$$

LeRoy et al. (1981) also used the Rice–Tracey model. They assumed that the void radii in the principal directions, R_1 and R_3, are related to their rates of increase as follows:

$$\left.\begin{array}{l}\dot{R}_1 = R_1(\tfrac{1}{2}\gamma_a\dot{\varepsilon}_3^\infty + D\dot{\varepsilon}_3^\infty), \\[4pt] \dot{R}_3 = R_3(\gamma_a\dot{\varepsilon}_3^\infty + D\dot{\varepsilon}_3^\infty), \end{array}\right\} \tag{5.31}$$

where $D = 0.56 \sinh(3\sigma_m/2\sigma_y)$, σ_y being the yield stress; γ_a is a factor describing the amplification of the growth rate of the void relative to the strain rate of the matrix. Assuming that an isolated void can be idealized to be an oblate ellipsoid, which can be characterized by its principal radii of curvature, then a simple geometrical condition for void interlinkage is when the longest axis $2R_3$ is of the order of the mean planar nearest-neighbour spacing λ:

$$2R_3^f = \phi\lambda, \tag{5.32}$$

where ϕ is assumed to be a constant for a given material. This type of interlinkage is very similar to that assumed by Thomason (1968). One feature of this analysis that may lead to error is the use of the Rice–Tracey model just prior to interlinkage—nothing is mentioned by Rice and Tracey about void interlinkage, because they were concerned with an isolated void. Brownrigg et al. (1983) studied the effect of hydrostatic pressure on the tensile ductility of a steel. Starting from the equations that LeRoy et al. (1981) used for void growth in tension, Brownrigg et al. integrated this equation and linearized the sinh term to give

$$\frac{\varepsilon_f(p)}{\varepsilon_f(0)} = 1 + \frac{1.68p}{2\bar{\sigma}}, \qquad (5.33)$$

where $\varepsilon_f(p)$ is the fracture strain under hydrostatic pressure and $\varepsilon_f(0)$ is the fracture strain at atmospheric pressure. Figure 5.13 shows good agreement between the experiments of Brownrigg *et al.* and the theory.

Brownrigg *et al.* also measured the fracture surface dimple size as a function of hydrostatic pressure: their results are shown in Fig. 5.14. This figure clearly indicates a threshold hydrostatic pressure, in this case about 1500 MPa, above which the dimple size is zero. This seems to be the transition pressure above which shear fractures occur. This supposition is supported by detailed studies of fracture surfaces by Brownrigg *et al.*

5.2.3 Porous plasticity and Gurson's continuum model

Oyane (1972) was of the opinion that ductile fracture occurs when the volume fraction of voids reaches a characteristic value. He used porous-plasticity theory in the form of a modified von Mises material in which

$$\left. \begin{array}{l} d\varepsilon_1 - \dfrac{d\varepsilon_v}{3} = \dfrac{d\bar{\varepsilon}}{\gamma\bar{\sigma}}\left(\sigma_1 - \dfrac{\sigma_2}{2} - \dfrac{\sigma_3}{2}\right), \\[2ex] d\varepsilon_2 - \dfrac{d\varepsilon_v}{3} = \dfrac{d\bar{\varepsilon}}{\gamma\bar{\sigma}}\left(\sigma_2 - \dfrac{\sigma_3}{2} - \dfrac{\sigma_1}{2}\right), \\[2ex] d\varepsilon_3 - \dfrac{d\varepsilon_v}{3} = \dfrac{d\bar{\varepsilon}}{\gamma\bar{\sigma}}\left(\sigma_3 - \dfrac{\sigma_2}{2} - \dfrac{\sigma_1}{2}\right), \end{array} \right\} \qquad (5.34)$$

and

$$d\varepsilon_v = \frac{d\bar{\varepsilon}}{\gamma f^2}\left(\frac{\sigma_m}{\bar{\sigma}} + a_0\right), \qquad (5.35)$$

where γ is the relative density, f is some function of γ, $d\varepsilon_v$ is the volumetric strain increment and a_0 is a constant. Assuming that fracture occurs at a volumetric strain of ε_{vf}, the fracture strain $\bar{\varepsilon}_f$ is given by integrating (5.35). Then for a solid material in which voids are nucleated after a certain critical strain,

$$\int_{\varepsilon_i}^{\varepsilon_{vf}} \frac{\gamma f^2}{a_0}\, d\varepsilon_v = \int_{\varepsilon_i}^{\bar{\varepsilon}_f}\left(1 + \frac{1}{a_0}\frac{\sigma_m}{\bar{\sigma}}\right) d\bar{\varepsilon}. \qquad (5.36)$$

Gurson (1977) proposed a continuum model of porous plasticity based on the analysis of the plastic stress field in a material that contains voids. The

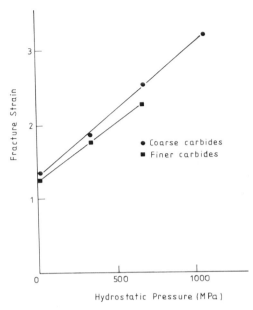

Fig. 5.13 Fracture strain versus hydrostatic pressure for both coarse and fine carbides. (After A. Brownrigg *et al.*, *Acta Metall.* **31**, 1141–1150 (1983).)

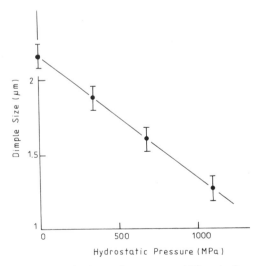

Fig. 5.14 Dimple size versus hydrostatic pressure. (After A. Brownrigg *et al.*, *Acta Metall.* **31**, 1141–1150 (1983).)

objective of Gurson's model is to show how voids grow during plastic deformation and its effect on the ductility of the material.

The yield function in Gurson's model is (for the derivation see Appendix 5B)

$$
\left.
\begin{aligned}
F(\Sigma_m, \bar{\Sigma}) = \Phi(\Sigma_{ij}, \bar{\sigma}, f) = \left(\frac{\bar{\Sigma}}{\bar{\sigma}}\right)^2 + 2f\cosh\left(\frac{3}{2}\frac{\Sigma_m}{\bar{\sigma}}\right), \\
-(1 + f^2) = 0,
\end{aligned}
\right\}
\tag{5.37}
$$

where Σ_{ij} denotes the macroscopic stress of the porous medium, $\bar{\Sigma}$ and Σ_m are the corresponding effective stress and mean stress, f is the volume fraction of voids and $\bar{\sigma}$ is the effective stress of the matrix. Clearly, (5.37) shows the relationship between $\bar{\Sigma}$ and Σ_m, the two invariants of the macroscopic stress, with parameters $\bar{\sigma}$ and f.

The complete description of the deformation of a porous material by plastic deformation in terms of Gurson's model consists of the stress–strain relation due to normality:

$$
\dot{E}_{ij} = \Lambda \frac{\partial \Phi}{\partial \Sigma_{ij}},
\tag{5.38}
$$

where Λ is determined by strain hardening and E_{ij} is the macroscopic strain of the porous medium. Also,

$$
\Sigma_{ij}\dot{E}_{ij} = (1 - f)\bar{\sigma}\dot{\bar{\varepsilon}},
\tag{5.39}
$$

according to the plastic work consumed, and

$$
\dot{f} = (1 - f)\dot{E}_{kk},
\tag{5.40}
$$

which comes from simple geometrical considerations.

Hence (5.37)–(5.40) constitute a complete theory of plastic deformation with voids. Although this is phenomenological, it is based on physical modelling of voids. In addition, (5.37) reduces to J_2 flow theory; that is, a von Mises yield function, when $f = 0$ (this is trivial) and $\Sigma_m = 0$. The latter corresponds to

$$
\bar{\Sigma} = (1 - f)\bar{\sigma},
\tag{5.41}
$$

which is a version of the von Mises function with a modification allowing for a void volume fraction. These factors are the advantages of Gurson's model over others.

Void nucleation can also be incorporated into the model (see the work of Saje *et al.*, 1982):

$$
\dot{f}_{\text{nucleation}} = A\dot{\bar{\sigma}} + B\dot{\Sigma}_m,
\tag{5.42}
$$

where A and B are two parameters representing the dependence of nucleation on the rates of effective stress of the matrix and the hydrostatic stress respectively.

Since voids perhaps are the trigger for plastic shear localization, a number of researches have been carried out with Gurson's model, dealing with the onset of localization as a bifurcation of the band form of intensified deformation from a homogeneous deformation mode. To fit relevant data, Tvergaard (1981) proposed a modified Gurson model with several constants:

$$\Phi(\Sigma_{ij}, \bar{\sigma}, f) = \left(\frac{\Sigma}{\bar{\sigma}}\right)^2 + 2fq_1\left(\frac{3}{2}q_2\frac{\Sigma_m}{\bar{\sigma}}\right) - (1 + q_3 f^2) = 0 \qquad (5.43)$$

where q_1, q_2 and q_3 are constants. Based on a study of the onset of localization, Tvergaard suggested that

$$q_1 = 1.5, \quad q_2 = 1, \quad q_3 = q_1^2.$$

5.2.4   Some remarks on theories

The theories concerning void growth, as well as nucleation, can almost always be written in terms of a critical strain, which is usually a fairly strongly decreasing function of hydrostatic stress. On the other hand, all of the above theories of void growth suffer from one major deficiency; namely, although they describe void growth fairly accurately, if they consider coalescence, it is always by internal necking of the ligaments between cavities. Another way of regarding the problem is that, after some critical event such as the attainment of a particular strain or a critical intervoid spacing, the ligaments neck down until a series of internal ruptures occur. The prevailing evidence, however, from innumerable experimental studies, is that at some stage in the fracture process localized shear occurs across the ligaments. Within the shear bands themselves, it is still possible (but not essential) that a second series of voids can nucleate. Shear banding of this type is discussed in detail in Chapters 6 and 7.

A further disadvantage of many of the analyses is that they only take into account the dilatational effect on voids, except for the formal Rice–Tracey approach. Interestingly, the velocity field presented by Rice and Tracey does include a term that is the shape-change contribution rather than due to dilatational effects, but, because this term is apparently insignificant in tension, it is neglected by Hancock and Mackenzie.

In addition, in recent years, the theories based on porous plasticity, such as Oyane's and, in particular, Gurson's models, have been quite widely adopted in the study of shear localization, another basic mechanism of ductile fracture. Hence the interaction of voids and localized plastic deformation, as happens in practice, can be revealed.

Appendix 5A

1   A circular hole in a non-strain-hardening von Mises material deforming in plane strain with a superimposed homogeneous increment of strain $d\varepsilon_z$

In an axisymmetrical stress state there is only one independent variable, r. Incompressibility requires $d\varepsilon_r + d\varepsilon_\theta + d\varepsilon_z = 0$, i.e.

$$\frac{\partial(du)}{\partial r} + \frac{du}{r} = -d\varepsilon_z, \tag{A1}$$

where all strains are logarithmic, $d\varepsilon_z = dl/l$, $d\varepsilon_r = \partial(du)/\partial r$ and $d\varepsilon_\theta = du/r$, and $u = u(r)$ is the displacement along the r-axis. Integration of (A1) gives

$$\left.\begin{aligned}
d\varepsilon_r &= -\frac{b^2}{r^2}\left(d\varepsilon_{\theta b} + \frac{d\varepsilon_z}{2}\right) - \frac{d\varepsilon_z}{2}, \\
d\varepsilon_\theta &= \frac{b^2}{r^2}\left(d\varepsilon_{\theta b} + \frac{d\varepsilon_z}{2}\right) - \frac{d\varepsilon_z}{2},
\end{aligned}\right\} \tag{A2}$$

where $d\varepsilon_{\theta b} = d\varepsilon_\theta|_{r=b}$. Now, a non-strain-hardening von Mises material is described by $\bar\sigma = $ constant and

$$d\varepsilon_\theta = \tfrac{3}{2}(\sigma_\theta - \sigma_m)\frac{d\bar\varepsilon}{\bar\sigma} \quad \text{etc.,} \tag{A3}$$

where $\bar\sigma$ and $d\bar\varepsilon$ are the effective stress and strain increment respectively. The equilibrium equation requires

$$\frac{d\sigma_r}{dr} + \frac{\sigma_r - \sigma_\theta}{r} = 0. \tag{A4}$$

Substitution of (A4) into (A3) gives

$$-r\frac{d\sigma_r}{dr} = \frac{2\bar\sigma}{3\,d\bar\varepsilon}(d\varepsilon_r - d\varepsilon_\theta). \tag{A5}$$

Differentiation of (A2) leads to

$$-r\frac{d\varepsilon_r}{dr} - \frac{2b^2}{r^2}\left(d\varepsilon_{\theta b} + \frac{d\varepsilon_z}{2}\right) = d\varepsilon_r - d\varepsilon_\theta. \tag{A6}$$

Substitution of (A6) into (A5) gives

$$\frac{1}{\bar\sigma}\frac{d\sigma_r}{d\varepsilon_r} = \frac{2}{3}\frac{1}{d\bar\varepsilon}. \tag{A7}$$

Recalling the definition of $d\bar{\varepsilon}$ and integrating (A7), one obtains

$$\sqrt{3}\frac{\sigma_r}{\bar{\sigma}} = \int \frac{2}{3}\frac{d\varepsilon_r}{d\bar{\varepsilon}} = \int \frac{d\varepsilon_r}{(d\varepsilon_r^2 + d\varepsilon_r\,d\varepsilon_z + d\varepsilon_z^2)^{1/2}}$$

$$= \sinh^{-1}\left(\frac{d\varepsilon_r + \frac{1}{2}d\varepsilon_z}{\frac{1}{2}\sqrt{3}\,d\varepsilon_z}\right) + C. \tag{A8}$$

At $r = \infty$

$$d\varepsilon_{r\infty} = d\varepsilon_{\theta\infty} = -\tfrac{1}{2}d\varepsilon_z, \quad \sigma_{r\infty} = \sigma_{\theta\infty}, \quad \bar{\sigma} = |\sigma_{z\infty} - \sigma_{r\infty}|.$$

Hence

$$d\varepsilon_\theta = -d\varepsilon_r - d\varepsilon_z = \sqrt{3}\,d\varepsilon_{r\infty}\sinh\left\{\frac{\sqrt{3}\,(\sigma_r - \sigma_{r\infty})}{\bar{\sigma}}\right\} + d\varepsilon_{r\infty}. \tag{A9}$$

Now at $r = R$

$$d\varepsilon_{\theta b} = d\left(\ln\left(\frac{R}{R_0}\right)\right) \quad \text{and} \quad \sigma_r|_{r=R} = 0,$$

which, together with (A9) lead to (5.12).

2 Extrapolation to an elliptical hole in the same material

Although comparison of circular and elliptical cases in a viscous medium gives a clearer idea of the substitution of parameters, because of limitations of space the following only concerns plastic media.

Substitutions from the case of a circular hole (equation 5.12) to an elliptical hole are

	circular		elliptical		giving equation
	$\sigma_{r\infty}$	$\rightarrow$	$\frac{1}{2}(\sigma_a + \sigma_b)$		(A10)
	$\sigma_{r\infty} - \sigma_{z\infty}$	$\rightarrow$	$\bar{\sigma}$		(A11)
first	$d\varepsilon_{r\infty}$	$\rightarrow$	$\frac{1}{2}d\bar{\varepsilon}$		(A12)
second	$d\varepsilon_{r\infty}$	$\rightarrow$	$\frac{1}{2}(d\varepsilon_a + d\varepsilon_b)$		(A13)

$d\varepsilon_{r\infty}$ is split into two effects. In the case of (A12), because of a plane-strain stress state, $d\varepsilon_a + d\varepsilon_b = 0$, which can lead to $d\varepsilon_{\theta b} = 0$. The second effect is that of the superimposed homogeneous increment of strain, $d\varepsilon_z$; hence $d\varepsilon_{r\infty} \rightarrow \frac{1}{2}(d\varepsilon_a + d\varepsilon_b)$.

3 Extrapolation to a strain-hardening material

For a strain-hardening material $0 < n < 1$. In the limit $n = 1$ there is a linear relation between stress and strain. Since strain hardening retards void growth, McClintock incorporated n as an interpolator to express the hardening effect as $1 - n$ in (5.17). As a check, at $n = 0$, (5.17) degenerates to (5.14). At $n = 1$ a first-order expansion of (5.17) gives

$$\ln\left(\frac{R}{R_0}\right) \sim \frac{\bar{\varepsilon}\sqrt{3}}{2(1-n)}\left\{\frac{\sqrt{3}}{2}\frac{(1-n)(\sigma_a + \sigma_b)}{\bar{\sigma}}\right\} + \frac{\varepsilon_a + \varepsilon_b}{2}$$

$$= \frac{3\bar{\varepsilon}}{4}\frac{(\sigma_a + \sigma_b)}{\bar{\sigma}} + \frac{\varepsilon_a + \varepsilon_b}{2},$$

which coincides with the solution for a linear material.

Appendix 5B

In Gurson's model the medium containing random voids is idealized as a single void in a solid matrix. The matrix is considered to be homogeneous, incompressible, and a rigid perfectly plastic material in the fully plastic and axisymmetrical deformation mode. Hence, similarly to (A3) and (A4), the strain rates in the matrix $\dot{\varepsilon}_{ij}$ are

$$\left.\begin{array}{l}
\dot{\varepsilon}_{rr} = -\dfrac{\dot{E}}{r^2} b^2 - \dfrac{\dot{E}_z}{2}, \\[3mm]
\dot{\varepsilon}_{\theta\theta} = \dfrac{\dot{E}}{r^2} b^2 - \dfrac{\dot{E}_z}{2}, \\[3mm]
\dot{\varepsilon}_{zz} = \dot{E}_z,
\end{array}\right\} \tag{B1}$$

where $\dot{E}$ and $\dot{E}_z$ are constants.

From the energy point of view, the macroscopic stress Σ_{ij} is given by

$$\Sigma_{ij} = \frac{\partial \dot{W}}{\partial \dot{E}_{ij}} = \frac{1}{V}\int_V s_{kl}\frac{\partial \dot{\varepsilon}_{kl}}{\partial \dot{E}_{ij}}\,dV = \frac{1}{V}\int_V \bar{\sigma}\frac{\partial \dot{\bar{\varepsilon}}}{\partial \dot{E}_{ij}}\,dV, \tag{B2}$$

where s_{ij} and $\dot{\varepsilon}_{ij}$ are the microscopic stress deviator and strain rate of the matrix respectively. Recalling that

$$\dot{\bar{\varepsilon}} = (\tfrac{2}{3}\dot{\varepsilon}_{kl}\dot{\varepsilon}_{kl})^{1/2}$$

and that $\bar{\sigma}$ is a constant for a perfectly plastic medium, (B2) becomes

$$\frac{\Sigma_{ij}}{\bar{\sigma}} = \frac{1}{V}\int_V \frac{\dot{\varepsilon}_{kl}}{\dot{\bar{\varepsilon}}}\frac{\partial \dot{\varepsilon}_{kl}}{\partial \dot{E}_{ij}}\,dV. \tag{B3}$$

Substitution of (B1) into the definition of $\dot{\bar{\varepsilon}}$ gives

$$\dot{\bar{\varepsilon}} = \sqrt{\frac{2}{3}\left\{2\left(\frac{\dot{E}b^2}{r^2}\right)^2 + \frac{3}{2}\dot{E}_z^2\right\}^{1/2}} = \frac{2}{\sqrt{3}}\left\{\frac{\dot{E}^2b^4}{r^4} + \frac{3}{4}\dot{E}_z^2\right\}^{1/2}. \tag{B4}$$

The definition of the macroscopic strain rate is

$$\dot{E}_{ij} = \frac{1}{V}\int_V \dot{\varepsilon}_{ij}\,\mathrm{d}V, \tag{B5}$$

where $V = \pi b^2 L$ and $\mathrm{d}V = 2\pi L r\,\mathrm{d}r$. After substituting (B1), (B4) and (B5) into (B3), one obtains

$$\left.\begin{aligned}
\frac{\bar{\Sigma}}{\bar{\sigma}} &= \frac{\Sigma_{33} - \Sigma_{11}}{\bar{\sigma}} = \frac{\sqrt{3}}{2}\frac{1}{V}\int_V \dot{E}_z\left(\frac{\dot{E}^2b^4}{r^4} + \frac{3}{4}\dot{E}_z^2\right)^{-1/2}\mathrm{d}V, \\[2mm]
\frac{\Sigma_{rr}}{\bar{\sigma}} &= \frac{2\Sigma_{11}}{\bar{\sigma}} = \frac{2}{\sqrt{3}}\frac{1}{V}\int_V \frac{\dot{E}b^4}{r^4}\left(\frac{\dot{E}^2b^4}{r^4} + \frac{3}{4}\dot{E}_z^2\right)^{-1/2}\mathrm{d}V.
\end{aligned}\right\} \tag{B6}$$

Noticing that the integration covers $r = (a, b)$ and letting $x = \dot{E}b^2/r^2$, $g = \frac{3}{4}\dot{E}_z^2$ (this is a similar transformation of variables to that of A11), (B6) can be written and integrated as

$$\left.\begin{aligned}
\frac{\bar{\Sigma}}{\bar{\sigma}} &= \frac{\sqrt{3}}{2}\int_{\dot{E}}^{\dot{E}/f} x^{-2}(x^2 + g)^{-1/2}\,\mathrm{d}x = g^{-1/2}\{(\dot{E}^2 + g)^{1/2} - (\dot{E}^2 + gf^2)^{1/2}\}, \\[2mm]
\frac{\Sigma_{rr}}{\bar{\sigma}} &= \frac{2}{\sqrt{3}}\int_{\dot{E}}^{\dot{E}/f} (x^2 + g)^{-1/2}\,\mathrm{d}x = \frac{2}{\sqrt{3}}\left\{\sinh\left(\frac{\dot{E}}{g^{1/2}f}\right) - \sinh^{-1}\left(\frac{\dot{E}}{g^{1/2}}\right)\right\}.
\end{aligned}\right\} \tag{B7}$$

Hence

$$\cosh\left(\frac{\sqrt{3}}{2}\frac{\Sigma_{rr}}{\bar{\sigma}}\right) = \left\{1 + f^2 - \left(\frac{\bar{\Sigma}}{\bar{\sigma}}\right)^2\right\}\Big/2f. \tag{B8}$$

In the case of spherical symmetry, $3\Sigma_m$ would appear in (B8) instead of $\sqrt{3}\,\Sigma_{rr}$.

6

Adiabatic Shear

When a metal is plastically deformed a small fraction of the work of plastic deformation is retained in the metal. This fraction is called the latent energy of cold work and is associated with the new defect structure of the metal. The remaining proportion of the work of plastic deformation, typically about 90% of the total, is converted into heat. Because the flow stress of metals decreases as the temperature increases, under conditions of large strain and high strain rate it is possible for thermally weakened zones or bands to develop. These thermal weakening effects can be accentuated by a tooling-workpiece geometry that itself requires extensive localized shearing. Once localized shearing begins, it can develop into the formation of adiabatic shear bands and eventually sudden, or catastrophic, fracture. Hence adiabatic shear is a typical case of shearing, being governed by material properties.

An adiabatic process is defined as one in which no heat is lost to the surroundings. Clearly, adiabatic plastic deformation is approached when the process of plastic deformation is carried out at such a high strain rate that the heat produced does not have the time to be conducted away from the deformation zone.

6.1 Observations of adiabatic shear

Adiabatic shear has been observed in such processes as armour penetration, dynamic blanking, high-speed machining and forming, fragmentation of cylindrical shells and bombs, rubbing surfaces and cryogenic plastic deformation. In many of these processes the common characteristic is localization of plastic deformation into well-defined shear zones.

Friction between the tooling and workpiece plays a key role in determining the mode of plastic deformation. An illustration of this is upset forging of a cylindrical workpiece. When good lubrication is used the cylinder is compressed uniformly into a flat disc. However, when the compression is carried out without lubrication nonuniform deformation occurs and the cylinder becomes barrelled markedly. If a barrelled cylinder is sectioned and etched dead-metal zones are observed that have adhered to the dies, and two bands of intense shear deformation, which cross in the middle of the specimen, are seen.

High-speed processes that confine plasticity to a small zone, for example machining, can produce a very high strain rate in the zone, since the average strain rate is defined as the process velocity divided by the width of the plastic zone. As we have observed, since the heat produced requires a finite time to be conducted away, the higher the strain rate the less heat is conducted away. It is important to appreciate that a process geometry that confines plastic flow is not a prerequisite for localized shear. This is underlined by the observations of localized adiabatic and nonadiabatic shearing in high-speed torsion tests on thin-walled tubes.

6.1.1 Projectile impact and dynamic punching

In early experiments on punching, Zerner and Hollomon (1944) observed the effects of localized shear deformation. The apparatus that was used is shown schematically in Fig. 6.1. A weight is dropped onto the buffer plate, and the plunger perforates the sheet, forcing a plug of material ahead of its nose. The sheet material was a carbon steel. When a slow pressure is applied to the buffer plate, the plastic deformation is not localized. However, in dynamic testing the plastic deformation is highly localized. A simple explanation of the observations made in dynamic loading may be given. On impact, plastic deformation is confined to the edge of the punch. The strain and strain rate in this region rapidly become large, and this is accompanied by a rise in the local temperature. The temperature rise further weakens the material because of thermal softening, and, as the punch advances, a shear band propagates to the lower surface of the plate and a plug forms by catastrophic shear fracture.

On microexamination of a partially sheared-out plug, the material in the shear zone contained white etching bands when etched with nital. From hardness measurements it was concluded that this white etching constituent consisted of martensite. Zerner and Hollomon estimated that the thin band of material had undergone a shear strain of about 100. Also, it was shown that a shear strain of 5 would have produced a local temperature rise in excess of 1000 °C. In projectile-impact studies, temperatures of this order are produced

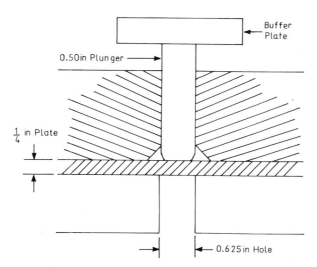

Buffer
Plate

0.50in Plunger

$\frac{1}{4}$ in Plate

0.625in Hole

Fig. 6.1 Apparatus for plugging experiments used by C. Zener and J. H. Hollomon (*J. Appl. Phys.* **15**, 22–32 (1944)).

in a matter of microseconds or milliseconds, and then the band is quenched by the surrounding bulk of the material.

Similar results to those of Zener and Hollomon have been observed in numerous projectile-impact studies. Considering the normal impact by a flat-nosed cylindrical projectile against a rigidly supported circular plate, the kinetic energy of the projectile can be absorbed by four basically different mechanisms, depending on the plate thickness, the projectile velocity and the plate and projectile materials. These mechanisms of energy absorption are membrane stretching of the target, formation and travel of plastic hinges in the target, perforation of the target (plugging) and mushrooming or fragmenting of the projectile (see Johnson, 1972a). It is found that there is a critical impact velocity at which plugging occurs. The energy absorbed in the plugging process is, in general, much smaller than that absorbed in either stretching or bending of the target.

Zener (1948), assuming that the planes of intense shear will coincide with the maximum shear stress planes before strain localization, showed qualitatively the effects of a superimposed tension and compression on plug shape (see Fig. 6.2). For pure plugging, the plane of strain localization will be normal to the plate thickness, as in Fig. 6.2(a). However, when a tension or compression is superimposed on this shear deformation, truncated plugs result, as shown in Figs. 6.2(d) and (e) respectively. A superimposed radial tension is equivalent to membrane stretching of the target. If the target is

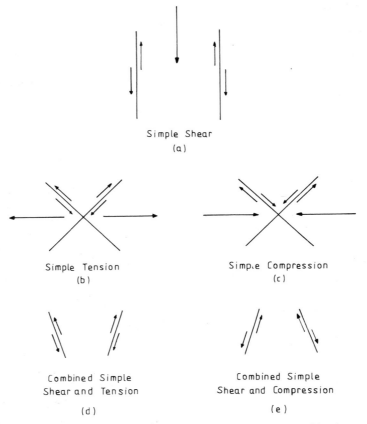

Fig. 6.2 Orientation of planes of maximum shear stress under combined stress states: (a) simple shear; (b) simple tension; (c) simple compression; (d) combined shear and tension; (e) combined shear and compression. (After C. Zener, *Fracturing of Metals*, pp. 3–31. ASM, Ohio, 1948.)

subjected to bending then the plug will be truncated. Depending on the sign of the bending moment, the plug should start truncated and end as an inverse truncated one, or vice versa.

It has been found by numerous researchers that the adiabatic shear bands formed in projectile impact studies are somewhat similar to those formed in dynamic blanking. This has led to the use of the blanking test to simulate the plugging phenomenon. For the study of adiabatic shear bands the blanking test is attractive; however, to use it for the study of plugging in projectile impact is fraught with dangers, since the structural response in the two processes is different.

6.1.2 Machining and forming processes

Adiabatic shear deformation may be approached in high-speed machining. The geometry of the machining process provides severe localization of deformation into small zones.

For a titanium workpiece Recht (1964) predicted a shear-zone temperature of 650 °C at a machining speed of 43 m min^{-1}. Further studies indicate that adiabatic shearing is of importance in the process of chip formation in the machining of many different materials over a range of machining speeds.

Dao and Shockey (1979) have recently measured temperatures in orthogonal machining of 4340 steel and 2016-T6 aluminium using an infrared microscope and a light-duty lathe. The shear bands in the cut steel between the chip segments etched white and were very hard, indicating that they were martensitic. Figure 6.3(a) shows the well-defined periodic shear bands in the steel, and Fig. 6.3(b) shows a shear band in the aluminium. Peak temperatures of 180 °C for the steel and 100 °C for the aluminium were recorded. However, taking into account that the temperature measurements were made over an area greater than the width of the shear band, shear-band temperatures of 500 °C for the steel and 120 °C for the aluminium were estimated. It is observed in this work that, at high pressures, the temperature required for the formation of austenite may be significantly lower than that required in conventional heat treatment. Therefore the formation of martensite by sudden quenching of the shear bands is quite probable.

In metal-forming processes the process geometry can constrain the metal to flow through quite complex paths. This flow is by shear deformation, often with superimposed normal stresses. In compressive bulk-forming operations, particularly with high tool–workpiece interfacial friction, the magnitude of shear deformation can be large over relatively narrow regions, leading to high local strain rates, and forms what is known as a line of thermal discontinuity (Johnson, 1972a). In these processes, although the average strain rate may be quite low, the strain rate in these narrow regions of shear may be markedly higher. Therefore there is the possibility of the initiation of adiabatic or near-adiabatic shear bands in some of these processes. Whether the onset of adiabatic shear will develop fully into localized shear bands and fracture will depend on process geometry and the mechanical and physical properties of the workpiece. These aspects of forming processes will be considered in more detail in Chapter 7.

6.1.3 Explosive loading

It is not surprising that much of the information on explosive loading of cylinders and fragmentation of bombs remains classified. In the open

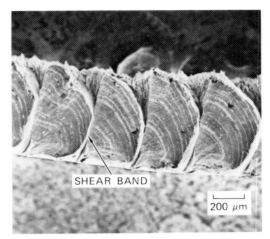

(a) 4340 Steel

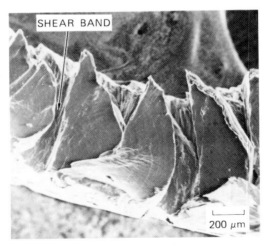

(b) 2014–T6 Aluminum

Fig. 6.3 Discontinuous chips in steel and aluminium. (After K. C. Dao and D. A. Shockey, *J. Appl. Phys.* **50**, 8244–8246 (1979).)

literature on this topic white etching bands have been observed on the fracture surfaces of fragments of steel cylinders and shells that have been internally explosively loaded. Also, white etching bands have been observed in explosively loaded thick-walled cylinders in which the strain rates have been estimated to be of the order of 10^7 s^{-1}.

In the process of explosive welding, shear bands are formed in the bonding area. The material within the shear bands is subjected to high strains and strain rates and high temperatures. With the help of transmission electron microscopy, it was shown by Hammerschmidt and Kerye (1981) that the bonding is achieved by a short-time melting of a narrow zone between about 0.5 and 5 μm in width. Another interesting observation is that two types of bands of intense shear appear in the deformed region. One is located in the bonded area and the other is inclined to the zone (Fig. 6.4). This is very similar to the two-shear-zone pattern in chip formation in high-speed machining.

6.2 Cryogenic plastic deformation

At cryogenic temperatures tensile deformation of a number of metals results in serrated stress–strain curves, which are similar to those observed due to dynamic strain ageing at elevated temperatures. Adiabatic heating is now accepted to be the cause of this phenomenon of serrated flow at cryogenic temperatures.

An example of jerky flow is shown in Fig. 6.5. It has been observed that the magnitudes of the discontinuities in the stress–strain curve increase with increasing strain, and their frequency increases with decreasing temperature. Also there is a threshold temperature above which flow is continuous.

The temperature rise due to a given amount of plastic work is dependent on the specific heat of the material. The specific heats of metals decrease with decreasing temperature, and therefore the temperature increase produced at cryogenic temperatures will be appreciably higher than that produced at room temperature for a given amount of energy. If the testing temperature is low enough then during testing the conditions may be adiabatic for some time. When the temperature increases in the specimen locally owing to the plastic work and the rate of thermal softening just exceeds the hardening effects of the strain and strain rate, flow will become unstable and the flow stress will decrease suddenly. However, the increase in specific heat due to temperature rise would decrease thermal softening. With continued straining the material will flow until the condition for adiabatic shear instability is reached again owing to the variations of temperature and specific heat. Basinski (1957) demonstrated both experimentally and theoretically that the temperature rise

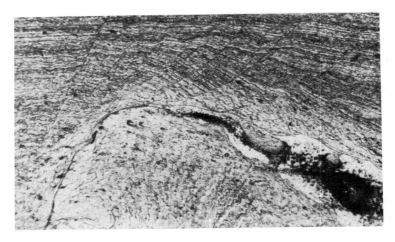

Fig. 6.4 Adiabatic shear bands in explosively welded steel. (Courtesy of Li Guohao, 1982.)

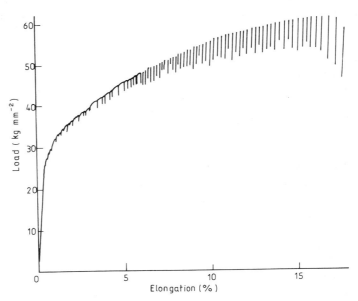

Fig. 6.5 A typical load–elongation curve at 4.2 K for solution-treated aluminium alloy. (After Z. S. Basinski, *Proc. R Soc. Lond.* **A240**, 229–242 (1957).)

and the time increment between instabilities are consistent with this argument.

It was observed by Rogers (1974) that the cryogenic deformation studies emphasize that adiabatic shear analyses define the conditions under which a strain perturbation or localization will be exaggerated. However, the heat generated just provides for a *possibility* of localization—an outside disturbance is required to nucleate the instability, as observed experimentally by Basinski (1957) and Chin *et al.* (1964a).

6.3 Nature of adiabatic shear bands and catastrophic fracture

Adiabatic shear bands have been observed in both ferrous and nonferrous metals, and they have been classified by Rogers (1974, 1979) as either deformation bands or transformation bands. Deformation bands are generally very narrow and there is no structural transformation within them. The majority of shear bands observed in nonferrous metals are of this type; an example is shown in Fig. 6.6. Transformation bands have been most commonly observed in ferrous alloys, in which they are associated with the ferrite–austenite transformation.

In steels that have been subjected to impact loading thin white etching lines have been observed in the shear bands. These white etching lines are thought to consist of very fine body-centred tetragonal martensite. It will be remembered that martensite is a nonequilibrium transformation product of face-centred cubic austenite. When a steel is heated to a high enough temperature, the body-centred cubic ferrite phase transforms to the higher-temperature austenite phase. If the austenite is quenched rapidly then the reverse transformation to ferrite is suppressed, and instead the nonequilibrium transformation product martensite is formed. Thus the process by which the fine white lines are formed is, first, localized adiabatic heating along the band, which forms a thin band of austenite, and secondly, rapid quenching of this austenite by the surrounding bulk of the material.

The martensite formed in high-strain-rate processes differs from martensite formed by conventional quenching. The major difference is that the martensite bands produced in high-rate processes have an equiaxed and finer grain size (typically tenths of microns) than conventional martensite. Further, from a limited number of studies, the martensite bands have, on occasion, been shown to contain retained austenite.

It is clear that the nature of the white etching bands will be strongly influenced by the composition and physical properties of the steel. The hardness of martensitic bands has been shown to increase with increasing projectile hardness in impact studies.

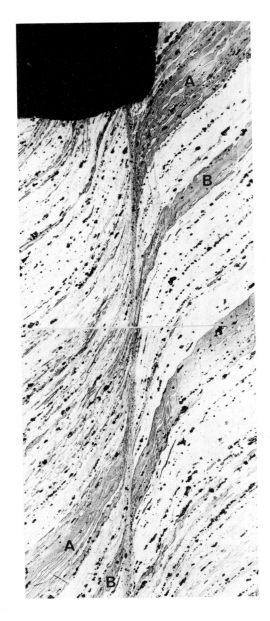

Fig. 6.6 A deformation band formed in an aluminium alloy beneath a flat-nosed projectile. The impact velocity of the projectile was 140 m s^{-1}. (After A. L. Wingrove, *Met. Trans.* **1**, 219–224 (1973).)

Apart from the above mentioned aspects relevant to the phase transformations, there are other features of shear bands that are worthy of note. Taking aluminium alloy 2014-T6 as an example, the shear strains may be as high as 100 in adiabatic shear bands, and these may be formed in less than 12 μs. Correspondingly, the shear strain rate is of the order of 10^7 s^{-1}, as described by Rogers (1974). Moss (1981) made a similar estimate for a nickel-chromium steel. Although there are lower estimates, for example those made by Olson *et al.* (1981), the estimates are still as high as 10 for the strain and about 10^5 s^{-1} for the strain rate. Also, estimates of shear-band widths for steels are about 10 μm. There is some evidence from scanning-electron-microscope observations to show that local melting may occur in the centre of shear bands, as described by Hartmann *et al.* (1981).

There is a tendency for pure metals with relatively high rates of strain hardening not to form adiabatic shear bands when subject to high-velocity deformation. Also, the number of observations of adiabatic shearing in metals subjected to pure tensile stress at high strain rate are minimal. For example, when uranium–2% molybdenum penetrators impact targets obliquely, shear bands are formed only in the region subjected to compression.

Catastrophic fracture is usually the result of the propagation of adiabatic shear bands. Because fractures occur in so many processes under different stress states and in a variety of metals, it is virtually impossible to say, from the examination of a fracture surface, whether or not fracture has been caused by adiabatic shear. Quite often adiabatic-shear fracture surfaces consist of parabolic dimple patterns indicative of shear in the material prior to and during separation. But equiaxed dimple patterns have also been observed. Equiaxed dimple patterns are caused by tensile stresses, indeed the fact that voids can nucleate and expand within a shear band shows that the band is not subjected to simple shear, at least during this period.

One feature that appears to be unique to some adiabatic shear fractures is the so-called "knobbly" appearance of the surface. This type of surface has been observed in aluminium alloys and occasionally in steels; an example is shown in Fig. 6.7. The accepted mechanism for the formation of this type of surface is that the heat generated locally by the rubbing surfaces of the matching fracture is so great that local melting occurs.

In alloy systems that exhibit transformation bands, once the transformation band has formed, because the product may be brittle, brittle fractures are observed. In these cases cracking occurs along the band or there are a number of separate cracks transverse to the band. The direction of propagation of the bands depends on the maximum shear directions.

Some examinations of fragmented bombs and impacted plates have shown that adiabatic shear bands have propagated along the maximum-shear-stress planes, and in bomb studies fracture also occurs along these planes.

Fig. 6.7 Knobbly fracture surface produced in steel. (After A. J. Bedford *et al.*, *J. Aust. Inst. Metals* **19**, 61–73 (1974).)

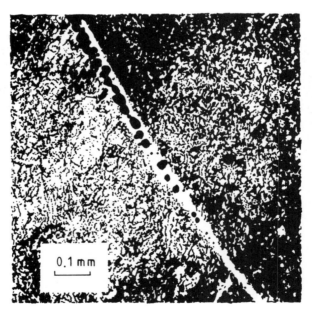

Fig. 6.8 Section of a fractured shear band in the nose of a uranium–molybdenum penetrator: as the band width decreases, the voids become more widely separated. (After C. J. Irwin, *Metallographic Interpretation of Impacted Ogive Penetrators*. DREV-R-652/72, Canada, 1972.)

Both brittle and ductile modes of fracture are observed in adiabatic shearing. Figure 6.8 shows a section of a shear band in a uranium–molybdenum penetrator. Void nucleation and coalescence occur within the thermally weakened band. But several distinct features of brittle fracture are also observed. Four different types of cracking are shown in Fig. 6.9. Microcracks can develop within adiabatic-shear bands or intersect them. The question of whether cracks initiate in transformed or untransformed material is still not answered satisfactorily.

As mentioned previously, adiabatic shearing caused by dynamic loading usually takes place in compressive stress fields, while cracks or voids are more probable under tensile stresses. These observations provide support for the general view of ductile fracture in combined stress states that is proposed here. Void initiation and growth and plastic shear instability are the two fundamental mechanisms of ductile fracture.

At the present time few attempts have been made to relate the morphology of fracture surfaces to the loading conditions, stress states and the planes along which adiabatic shear bands propagate. To understand fully the development of adiabatic shear bands and subsequent fractures is a very important field for current and future research.

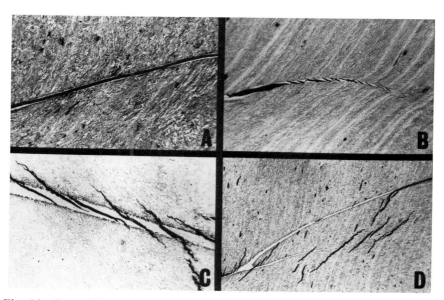

Fig. 6.9 Four different types of fracture associated with transformation adiabatic bands. (After M. E. Backman and S. A. Finnegan, in *Metallurgical Effects at High Strain Rates* (ed. R. W. Rohde *et al.*), pp. 531–543. Met. Soc. AIME/Plenum Press, New York, 1973.)

6.4 Simplified analysis of adiabatic and thermally assisted shear localization

6.4.1 Adiabatic-shear localization

It is logical to study localization of deformation in torsion tests, since adiabatic shear localization is being considered. The conventional constant-strain-rate nonisothermal stress–strain curve is depicted in Fig. 6.10(a). The maximum in the curve, corresponding to $d\tau = 0$, is the point at which shear strain begins to localize. Figure 6.10(b) shows the constant-strain-rate adiabatic stress–strain curve, in which, once again, there is a maximum, corresponding to $d\tau = 0$ and adiabatic shear localization. The isothermal constant-rate stress–strain curve does not exhibit a stress maximum, as shown in Fig. 6.10(c), and therefore shear localization is not possible in this case. Figures 6.10(d) and (e) illustrate the relationships between the flow stress and strain rate and temperature respectively. As the strain rate is increased, the flow stress at a given strain increases; this effect is known as strain-rate hardening. Also, as the temperature is increased the flow stress at a given strain decreases; this effect is known as thermal softening, as described previously.

If history effects are ignored as well as changes in the structure of the metal then the flow stress τ may be written as a function of the strain γ, the strain rate $\dot{\gamma}$ and the temperature T:

$$\tau = f(\gamma, \dot{\gamma}, T). \tag{6.1}$$

An increment in the flow stress $d\tau$ is given by

$$d\tau = \left(\frac{\partial \tau}{\partial \gamma}\right)_{\dot{\gamma}, T} d\gamma + \left(\frac{\partial \tau}{\partial \dot{\gamma}}\right)_{\gamma, T} d\dot{\gamma} + \left(\frac{\partial \tau}{\partial T}\right)_{\gamma, \dot{\gamma}} dT, \tag{6.2}$$

where the three differential quantities have the following straightforward meanings:

$\left(\dfrac{\partial \tau}{\partial \gamma}\right)_{\dot{\gamma}, T}$ is the isothermal rate of strain hardening at constant strain rate;

$\left(\dfrac{\partial \tau}{\partial \dot{\gamma}}\right)_{\gamma, T}$ is the isothermal rate of strain-rate hardening at constant strain;

$-\left(\dfrac{\partial \tau}{\partial T}\right)_{\gamma, \dot{\gamma}}$ is the rate of thermal softening at constant strain rate and strain.

Now at instability $d\tau = 0$, and if it is assumed that the plastic deformation

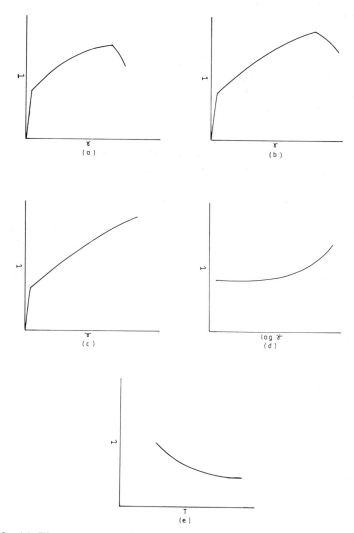

Fig. 6.10 (a) The constant-strain-rate nonisothermal stress–strain curve; (b) the constant-strain-rate adiabatic stress–strain curve; (c) the isothermal constant-rate stress–strain curve; (d) the common relationship between flow stress and the logarithm of the strain rate; (e) the typical relationship between flow stress and temperature (all schematic).

occurs at a constant strain rate for simplicity then (6.2) becomes

$$\left(\frac{\partial \tau}{\partial \gamma}\right)_{\dot{\gamma}, T} + \left(\frac{\partial \tau}{\partial T}\right)_{\gamma, \dot{\gamma}} \frac{\mathrm{d}T}{\mathrm{d}\gamma} = 0. \tag{6.3}$$

Remembering that the flow stress decreases as the temperature increases, i.e. $(\partial \tau / \partial T)_{\gamma, \dot{\gamma}}$ is negative, at instability

$$\left(\frac{\partial \tau}{\partial \gamma}\right)_{\dot{\gamma}, T} = \left|\left(\frac{\partial \tau}{\partial T}\right)_{\gamma, \dot{\gamma}}\right| \frac{\mathrm{d}T}{\mathrm{d}\gamma}. \tag{6.4}$$

This equation is the same as that derived by Recht (1964) in his considerations of high-speed machining.

For adiabatic deformation, the rate of increase of temperature with strain $\mathrm{d}T/\mathrm{d}\gamma$ is given by

$$\frac{\mathrm{d}T}{\mathrm{d}\gamma} = \frac{A\tau}{\rho c_v}, \tag{6.5}$$

where A is the fraction of the plastic work converted into heat ($A \approx 0.9$), τ is the flow stress, ρ is the density of the metal and c_v its specific heat. It should be noted that here the specific heat has been assumed not to vary with temperature; this is a good assumption for most metals at or above room temperature. Substituting (6.5) into (6.4) gives

$$\left(\frac{\partial \tau}{\partial \gamma}\right)_{\dot{\gamma}, T} = \frac{A\tau}{\rho c_v} \left|\left(\frac{\partial \tau}{\partial T}\right)_{\gamma, \dot{\gamma}}\right|. \tag{6.6}$$

From (6.6) it can be seen that a metal with a low density, low specific heat and a large temperature sensitivity of flow stress is more susceptible to adiabatic shear localization. Examples of metals that are susceptible to adiabatic shear therefore include aluminium and its alloys and titanium and its alloys.

If the metal is assumed to have an isothermal constant-strain-rate stress–strain curve of the power-law form $\tau = K\gamma^n$ then (6.6) becomes

$$\gamma_i = \frac{n\rho c_v}{A \left|\left(\frac{\partial \tau}{\partial T}\right)_{\gamma, \dot{\gamma}}\right|}. \tag{6.7}$$

The problem with applying (6.7) is that the isothermal stress–strain relation is difficult to determine, particularly at high strain rates. Instability strains have been calculated by Culver (1973) for three materials, these are as follows:

	n	γ_i
aluminium alloy	0.075	0.438
mild steel	0.28	1.94
titanium	0.17	0.326

From these values it is clear that the aluminium alloy and titanium can become unstable at small shear strains, whereas the steel can only become unstable at a relatively large strain. These results underline the effect of low values of densities and strain-hardening coefficients on the instability strain.

These findings are confirmed by the torsion experiments reported by Kudo and Tsubouchi (1971). These experiments were carried out on tapered tubular specimens, and it was found that the strain concentrated earlier for steel than for brass specimens. This behaviour can be explained by noting that these two materials had comparable n-values, but the thermal properties differed markedly.

In the case of cryogenic tensile tests, where the temperature sensitivity of the specific heat is of importance, it is necessary to determine the conditions for which $dL = 0$, where L is the load. It is permissible again to write the true stress σ as some function of the strain ε, strain rate $\dot{\varepsilon}$ and temperature T:

$$\sigma = f(\varepsilon, \dot{\varepsilon}, T). \tag{6.8}$$

An increment in stress $d\sigma$ is given by

$$d\sigma = \left(\frac{\partial \sigma}{\partial \varepsilon}\right)_{\dot{\varepsilon}, T} d\varepsilon + \left(\frac{\partial \sigma}{\partial \dot{\varepsilon}}\right)_{\varepsilon, T} d\dot{\varepsilon} + \left(\frac{\partial \sigma}{\partial T}\right)_{\dot{\varepsilon}, \varepsilon} dT. \tag{6.9}$$

Since $L = \sigma A$, $d\varepsilon = -dA/A$, where A is the cross-sectional area, and for simplicity a constant strain rate is assumed, (6.9) leads to

$$\left(\frac{\partial \sigma}{\partial \varepsilon}\right)_{\dot{\varepsilon}, T} + \left(\frac{\partial \sigma}{\partial T}\right)_{\varepsilon, \dot{\varepsilon}} \frac{dT}{d\varepsilon} - \sigma = 0 \tag{6.10}$$

at instability. $dT/d\varepsilon$ is given by

$$\frac{dT}{d\varepsilon} = \frac{A\sigma}{\rho c_v}; \tag{6.11}$$

then $dL = 0$ corresponds to

$$\left(\frac{\partial \sigma}{\partial \varepsilon}\right)_{\dot{\varepsilon}, T} = \sigma \left\{ \frac{A}{\rho c_v} \left| \left(\frac{\partial \sigma}{\partial T}\right)_{\varepsilon, \dot{\varepsilon}} \right| + 1 \right\}. \tag{6.12}$$

The difference between (6.6) and (6.12) is the presence of unity in the latter. This term reflects the importance of what is called geometrical softening. Geometrical softening is equivalent to a fluctuation in cross-sectional area.

If the isothermal constant-strain-rate stress–strain relation is assumed to be of the power-law form $\sigma = K\varepsilon^n$ then (6.12) gives

$$\varepsilon_i = \frac{n}{\dfrac{A}{\rho c_v} \left| \left(\dfrac{\partial \sigma}{\partial T}\right)_{\varepsilon, \dot{\varepsilon}} \right| + 1}. \tag{6.13}$$

6.4.2 Thermally assisted shear localization

In the discussion so far of adiabatic deformation, for simplicity all of the heat generated by the plastic work has been assumed to be confined to a narrow band. Within this narrow band shear localization occurs owing to the effect of thermal softening.

Figure 6.11 depicts a band of width a and length l that is undergoing shear deformation at a constant strain rate $\dot{\gamma}$. If the heat generated is wholly confined to the band then the temperature distribution for this adiabatic case is as indicated.

In most practical cases of plastic deformation there will always be some heat conducted away from the deforming zone. The amount of heat conducted away will depend upon the thermal conductivity of the metal and the local strain rate, among other variables. The resulting temperature distribution is shown in Fig. 6.11(b). For cases such as these, where conductivity cannot be ignored, this has led some authors to write in terms of an "almost-adiabatic process" or "semi-adiabatic process". More seriously, other authors term any process of plastic deformation in which thermal softening can lead to unstable flow as adiabatic. It is much less confusing and more accurate only to call processes in which no heat is lost to the surroundings adiabatic. The more general case in which, although thermal softening is important, conduction cannot be ignored, should be termed thermoplastic deformation. Then adiabatic shear localization or instability is a special case of thermoplastic shear localization.

A convenient and straightforward way to take into account this more general thermoplastic shear localization in the equations of the last section is to modify (6.5) and (6.6). Equation (6.5) becomes

$$\frac{\mathrm{d}T}{\mathrm{d}\gamma} = \frac{AB\tau}{\rho c_v}, \tag{6.14}$$

where $B = 1$ for adiabatic deformation and $B < 1$ for nonadiabatic deformation. From thermal considerations, B depends on the ratio $a/(\kappa t)^{1/2}$, where a is the width of the plastic zone, κ is the thermal diffusivity and t the time. Equation (6.6) becomes

$$\left(\frac{\partial \tau}{\partial \gamma}\right)_{\dot{\gamma}, T} = \frac{AB\tau}{\rho c_v}\left|\left(\frac{\partial \tau}{\partial T}\right)_{\gamma, \dot{\gamma}}\right|, \tag{6.15}$$

and it follows that if a power-law-hardening material is assumed then

$$\gamma_i = \frac{n\rho c_v}{AB\left|\left(\dfrac{\partial \tau}{\partial T}\right)_{\gamma, \dot{\gamma}}\right|}. \tag{6.16}$$

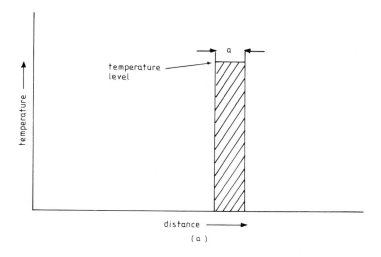

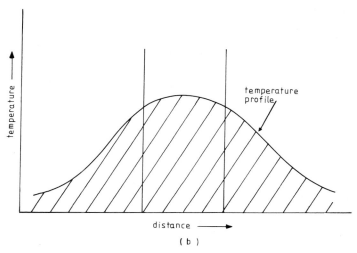

Fig. 6.11 (a) A narrow adiabatic shear band with the temperature rise confined to the width of the band; (b) a nonadiabatic shear band with a markedly greater width over which thermal softening can occur.

Clearly, as B decreases below unity, the instability strain becomes larger, and in the limit when $B \to 0$, $\gamma_i \to \infty$, i.e. for an isothermal process unstable flow in shear is impossible, ignoring the possibility of strain–rate softening, which occurs rarely.

Following the same procedure as with (6.13) leads to

$$\varepsilon_i = \frac{n}{\dfrac{AB}{\rho c_v} \left| \left(\dfrac{\partial \tau}{\partial T} \right)_{\varepsilon, \dot{\varepsilon}} \right| + 1}. \tag{6.17}$$

for uniaxial tension. For isothermal plastic deformation, $B = 0$, and the instability strain ε_i equals the strain-hardening coefficient n. This is the same result as given in (2.5).

It is, of course, possible to take into account the effects of strain-rate hardening in the derivation of the above equations, but this only leads to greater complexity.

6.4.3   Thermal and geometrical softening effects in shear

It is possible to allow for geometrical as well as thermal effects in shear deformation. Consider a shear band length l subjected to a shear force F. We assume that the band contains an array of equispaced voids of total width w along its length. Then the instability strain for a material that hardens isothermally according to the conventional power law is

$$\gamma_i = \frac{n(l - w)}{\mathrm{d}w/\mathrm{d}\gamma}. \tag{6.18}$$

If we assume that the stress and strain fields of the voids do not interact then, from (5.31), it is possible to assume that the void growth rate is of the form

$$\dot{w} = \dot{\gamma} \left(a + \frac{bp}{\sigma_y} \right), \tag{6.19}$$

where b is negative. If a void nucleation strain γ_c is assumed then (6.19) may be integrated directly to give

$$w = (\gamma - \gamma_c) \left(a + \frac{bp}{\sigma_y} \right) \quad \text{for } \gamma \geqslant \gamma_c. \tag{6.20}$$

This equation can be substituted into (6.18) to give the instability strain that can occur exclusively from geometrical effects.

To derive the corresponding expression for the instability strain under the combined effects of geometrical and thermal softening, we begin with

$$\gamma_i = \frac{n}{\dfrac{1}{M} + \dfrac{1}{l-w}\dfrac{dw}{d\gamma}} \quad \text{for } \gamma \geqslant \gamma_c, \qquad (6.21)$$

where

$$M = \frac{\rho c_v}{AB\left|\left(\dfrac{\partial \tau}{\partial T}\right)_{\gamma,\dot{\gamma}}\right|}$$

Substituting (6.19) into (6.21), we obtain the following implicit equation for the instability strain under the combined effects of geometrical and thermal softening:

$$\gamma_i = \frac{n}{\dfrac{1}{M} + \dfrac{a + bp/\sigma_y}{l - (\gamma_i - \gamma_c)(a + bp/\sigma_y)}} \quad \text{for } \gamma \geqslant \gamma_c. \qquad (6.22)$$

If the material is idealized such that it is assumed to contain no voids and no void-nucleation sites then the instability strain is given by (6.16). In the absence of thermal softening when p is compressive there is a threshold value of p/σ_y, given by $p/\sigma_y = -a/b$, above which instability due to geometrical effects is suppressed. Table 6.1 depicts the various possibilities for shear instability under compressive and tensile hydrostatic stresses. Material A is prone to thermoplastic shear localization, and the presence of a super-imposed hydrostatic stress will not change the instability strain. On the other hand, the onset of instability in material B is sensitive to hydrostatic stress. It is theoretically possible in a B-like material for voids to nucleate in the thermoplastic shear bands.

Using the above combined instability approach, it was shown by Dodd and Atkins (1983a) that the pressure sensitivity of the maxima in the shear-stress–strain curves obtained for steel by Osakada et al. (1977) can be explained (see §2.3.3).

6.5 Influence of system stiffness on adiabatic shear

In cryogenic tensile tests it has been observed that tapping the testing machine can initiate the load drop as discussed by Basinski (1957) and Chin et al. (1964b). Therefore the stiffness of the testing-machine–specimen system is a factor that must be considered in the development of unstable plastic flow under cryogenic conditions. Once initiated, the load will drop at a rate determined by the stiffness of the system.

Table 6.1

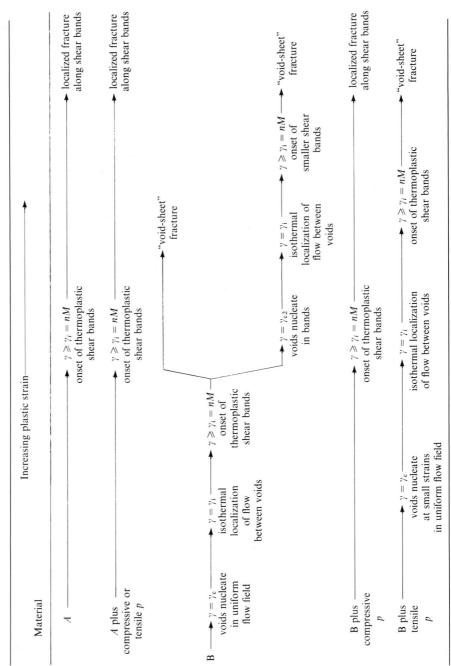

The effects of stiffness may be readily understood from the description of the load-drop phenomenon given by Chin *et al.* (1964b). Figure 6.12 shows a schematic load–elongation curve. Curve 1 is the isothermal curve at the cryogenic temperature, and curve 2 corresponds to higher temperatures produced by the plastic deformation. The specimen load–elongation relation follows the serrations. An unloading portion is governed by the system stiffness. Let a load drop occur at point A; the load will continue to decrease along ABD to point B. The discontinuous flow stops at B because the load required to deform the specimen along BC does not decrease. When arrested, the specimen cools to the testing temperature, so that elongation cannot continue along BC. The load must be raised to point E before flow can begin again. At point F elongation follows the system stiffness to G. Further elongation along GI is possible in this case, since from G onwards the load required to deform the specimen along GH is smaller than that produced by the stiffness of the system. Therefore, here, unstable flow leads to fracture.

The concepts of nucleating adiabatic shear by an outside disturbance and the importance of system stiffness may well be a common thread in many, if not all, examples of adiabatic deformation. It is only the rather unique conditions of cryogenic tensile tests that these effects are so clear.

A further example of machine stiffness effects has been described by Rogers (1974) in relation to machining experiments. It is found that localized shear

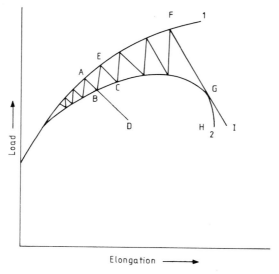

Fig. 6.12 Serrated load-elongation curve. (After G. Y. Chin *et al., Trans. Met. Soc. AIME* **230**, 1043–1048 (1964).)

bands are spaced uniformly along the chips. When the cutting speed increases, the shear-band spacing increases. Although Recht (1964) suggested differences in thermal gradients to explain the variation of spacing with cutting speed, Rogers has an alternative explanation. He suggests that the mass and elastic response of the system are the controlling features.

Lemaire and Backofen (1972) took into account the release of stored strain energy in machining, and they showed that plastic deformation alone is insufficient to raise the temperature of the shear zone to the austenite reversion temperature. However, taking account of the relaxation of stored energy in the shear zones leads to higher temperatures.

In the majority of processes, the elastic energy is either difficult to measure or it is masked by other effects. However, the way in which the strain energy is released can have a profound effect on localized shearing, as we have seen from the two examples. In general, once unstable flow is initiated, its duration is longer for a softer machine. Also, fracture is postponed to higher stresses with stiffer machines.

The physical and mechanical properties of a material that will influence the onset of adiabatic shear localization are a low density, specific heat and thermal conductivity, a low rate of strain hardening and strain-rate hardening, and a high temperature sensitivity of the flow stress.

A process geometry that requires highly localized deformation and a low stiffness of the loading system will increase the danger of adiabatic shear.

7

Theories of Ductile Fracture II: Localization of Plastic Deformation

It has been shown in numerous material tests and deformation processes that ductility is limited by strain localization. This highly localized deformation can take the form of a neck or a shear band. For example, a ductile thin strip exhibits localized plastic deformation in a visible narrow shear band—a local neck. Within a shear band it is possible for voids to nucleate, grow and eventually coalesce. Alternatively, shear bands can form between existing cavities and thereby weaken the ligaments (see McClintock, 1968; Rice, 1977). Generally, localized deformation is a precursor to ductile fracture. It has been observed that localization of plastic deformation occurs in single crystals, polycrystals and amorphous metals (see e.g. Peirce *et al.*, 1982; Steif *et al.*, 1982; Rice, 1977). This important feature would appear to imply that the microstructure of the material does not affect some general features of localization. However, microscopic inhomogeneities may play an extremely significant role in triggering the instability and related phenomena. These findings appear to favour the continuum-mechanics approach for a study of localization, but, as far as triggering is concerned, structural defects and other microscopic inhomogeneities cannot be neglected.

In the present chapter we shall clarify several of the concepts and localization phenomena of plastic deformation. We shall also discuss their physical significance and provide some of the mathematical tools for predicting the phenomena. Although the mathematics required is elegant, it is somewhat complicated. The emphasis here will be on physical understanding rather than mathematical manipulation. Furthermore, the effect of material behaviour rather than boundary conditions on localization is emphasized.

The interrelationship between localization and eventual ductile fracture, however, remains open.

7.1 Plastic localization: necking and shear banding

7.1.1 Diffuse necking

As has been discussed in Chapter 2, uniform deformation of a cylindrical bar or a strip pulled in tension eventually becomes localized, i.e. the specimen acquires a neck with a finite sharpness. The length of the neck is approximately equal to the thickness of the strip or the diameter of the bar. Both the initiation and the development of the neck are diffuse. Therefore the determination of the onset of diffuse necking is extremely difficult to ascertain both theoretically and experimentally. Diffuse necking is a progressive process.

7.1.2 Localized necking

A sheet specimen may thereafter undergo localized necking, which has also been described in Chapter 2. In localized necking all of the plastic deformation is concentrated in a narrow band with a fixed angle to the pulling axis. For localized necking to continue, the material within the neck must be subjected to the constraint of zero extension along its length.

7.1.3 Lüders bands

Lüders bands are a type of localized deformation (see Nadai, 1931; Cottrell, 1964; Johnson, 1972a), and their occurrence is closely related to discontinuous yielding shown in the stress–strain curve of Fig. 2.2. As the load reaches the upper yield point, for an annealed steel strip specimen a visible narrow band appears across the specimen. The material within the band is plastic, whereas the material on either side of the band remains elastic. The formation of a Lüders band requires inextensional deformation along the length of the band. Thus the angle with which the Lüders band is inclined to the loading axis is approximately 54.7°, the same as that for localized necking.

Unlike localized necking, the occurrence of Lüders bands is closely associated with discontinuous yielding. At either edge of a Lüders band a plastic wave front is present, although the velocity of propagation of the wave may be slow. The whole of the specimen becomes plastic at the end of the Lüders strain horizontal, point B in Fig. 2.2.

In summary a Lüders band appears at the upper yield point and disappears when the specimen is completely plastic. Clearly, Lüders bands constitute the transition between elastic and plastic regimes of deformation.

7.1.4 Slip-lines

It is well known that ductile fracture surfaces are often inclined to the principal stress directions and coincide with slip-lines (see e.g. Fig. 7.1).

Slip-lines are the names given to two families of characteristics along which inextensional shear occurs. However, velocity discontinuities are possible across the slip-lines. These properties can be illustrated clearly in the case of a rigid perfectly plastic material subjected to a plane-strain stress state. Because elastic strains are neglected, the Lévy–von Mises stress–strain rate equations can be written as

$$\frac{\dot{\varepsilon}_1}{\sigma_1 - \frac{1}{2}(\sigma_2 + \sigma_3)} = \frac{\dot{\varepsilon}_2}{\sigma_2 - \frac{1}{2}(\sigma_1 + \sigma_3)} = \frac{\dot{\varepsilon}_3}{\sigma_3 - \frac{1}{2}(\sigma_1 + \sigma_2)} = \text{constant}.$$

$$(7.1)$$

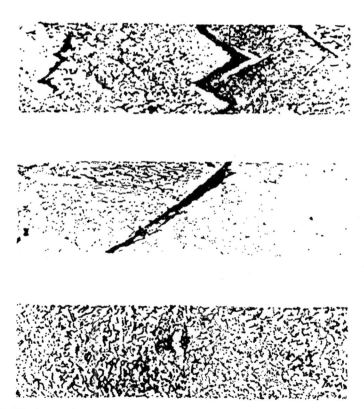

Fig. 7.1 Fracture along maximum-shear-stress planes. (After A. S. Korhonen (1981), from S. Kivivuori and M. Sulonen, *Ann. CIRP* **27**, 141–145 (1978).)

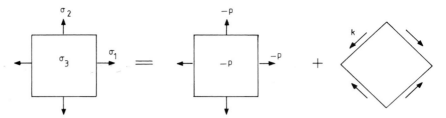

Fig. 7.2   Plane-strain state of stress in a plastic volume element.

Since $\dot{\varepsilon}_3$ is zero for plane-strain deformation,

$$
\left.
\begin{array}{l}
\sigma_3 = -p = \frac{1}{2}(\sigma_1 + \sigma_2), \\[4pt]
\sigma_2 = -p - \frac{1}{2}(\sigma_1 - \sigma_2), \\[4pt]
\sigma_1 = -p + \frac{1}{2}(\sigma_1 - \sigma_2),
\end{array}
\right\}
\tag{7.2}
$$

where σ_3 is obviously taken to be the intermediate principal stress with the value of $-p$. The stress state can be divided into a hydrostatic pressure p and a pure shear $k = \frac{1}{2}(\sigma_1 - \sigma_2)$, as shown in Fig. 7.2. The two families of characteristics referred to as slip-lines intersect orthogonally for both von Mises and Tresca solids. Here there is no normal strain along an element of slip-line (Fig. 7.3). However, a discontinuity in tangential velocity v is allowed, as long as the different values v_1 and v_2 satisfy the corresponding kinematic boundary condition on either side of the slip-line.

In the case of a rigid perfectly plastic solid a tangential velocity discontinuity is an approximation. In fact, slip-lines should be narrow transition zones within which intense shear occurs. These shear bands represent localized deformation and constitute some fracture patterns (Fig. 7.1).

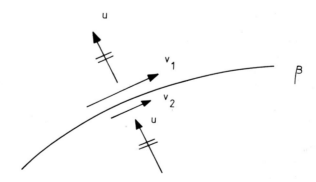

Fig. 7.3   Velocity difference allowed in the direction of slip-line.

7.1.5   Thermal shear bands

Examples of thermal shear localization have already been described in Chapter 6.

It is not really surprising that there is often a coincidence between slip-lines and thermal shear bands (Fig. 7.4), because thermal shear bands are a mode of shear-flow localization. A feature that makes thermal shear bands unique among modes of flow localization is that they are governed by a coupled thermoplastic mechanism.

7.1.6   General remarks on plastic localization

There are two basic forms of plastic localization, necking and shear banding, that have been recognized. These are different but interrelated processes.

Plastic flow localization can occur in tension and compression. Thinning always accompanies tension, while other modes of softening, such as the thermally assisted effect, may occur in compression.

Local necking and shear band formation are all limited to narrow regions. Because they cannot spread out in the direction of their normals, further localized plastic deformation can be separated from an external

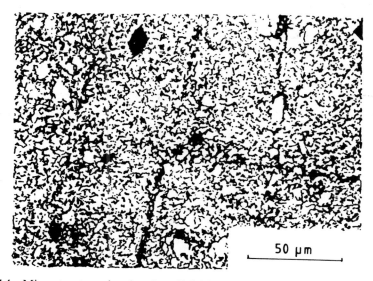

Fig. 7.4   Microstructure showing the adiabatic shear bands meeting at right angles that are similar to slip-lines. (After S. M. Doraivelu *et al.*, in *Shock Waves and High-Strain-Rate Phenomena*, (ed. M. A. Meyers and L. E. Murr), pp. 263–275. Plenum Press, New York, 1981.)

agency. One prerequisite is that no extension occurs along the weakened zone. For example, the two maximum shear directions satisfy the inextensional requirement in plane strain.

Localized plastic bands are almost always attributed to shear deformation, for example slip-lines and heat crosses. Even in local necking, which is thinning in tension, shear governs the process. The occurrence of localized bands under certain circumstances appears to be quite sensitive to sharp points or corners or pointed vertices on the yield surfaces, as the Lüders band indicates.

7.2   Criteria for localization of plastic flow

In attempts to identify the condition under which plastic flow localization occurs, a number of concepts have been proposed. These are necking (both diffuse and localized), shear bands, instability and bifurcation. As we have seen, both necking and shear banding are the major modes of plastic localization. Hereinafter plastic instability will be regarded as the point at which uniform deformation becomes unstable, i.e. a small additional disturbance would lead to a finite deviation from the original deformation path. By definition, bifurcation is neither instability nor the beginning of localization. A bifurcation occurs when the mode of deformation becomes nonunique. That is, at the bifurcation several solutions can satisfy the governing equations and boundary conditions; at this point it is possible for the material to deform in several different ways. Hence, at first sight, both instability and bifurcation do not imply localization definitely, and they are different concepts. Necking and localized shear banding are continuous modes of plastic deformation. The occurrence of either of them manifests a transition in deformation modes. In this sense there is a similarity between localization, bifurcation and instability. In addition, it is very difficult to pinpoint experimentally the beginning of diffuse necking, and therefore the onset of localized plastic flow is tied, theoretically, to bifurcation and instability as criteria.

This section is aimed at a straightforward mathematical and physical description of the transition to a localized mode of deformation and how that mode of localization develops, at least to a limited extent.

7.2.1   Physical outline

As soon as the effect of softening outweighs the strain-hardening effect, instability may occur, by definition. Nothing has been said about the process of localization itself. It is assumed that localization would appear first at a location where imperfections or disturbances are present.

Softening is usually the result of geometrical or material effects. The former is almost always responsible for necking. Material effects such as thermal softening can lead to adiabatic shear bands. Both strain and strain-rate hardening are examples of the hardening effect. In the case of a cylindrical rod the softening and hardening are due to dA and $d\sigma$ respectively. Because of incompressibility $\delta\varepsilon = -\delta A/A$, the load variation resulting from softening is given by

$$\frac{(\delta L)_S}{A\delta\varepsilon} = \frac{\sigma}{A}\frac{\delta A}{\delta\varepsilon} = -\sigma.$$

In a similar way

$$\frac{(\delta L)_H}{A\delta\varepsilon} = \frac{A}{A}\frac{\delta\sigma}{\delta\varepsilon} = \frac{d\sigma}{d\varepsilon}.$$

Balancing the two effects gives the critical condition for the onset of necking as

$$\frac{d\sigma}{d\varepsilon} = \sigma. \tag{7.3}$$

This balance approach has the advantage of a clear physical background and provides a good understanding of the phenomenon. However, it is not an accurate description of the process itself. Also the approach does not always work in complicated stress states.

7.2.2 Maximum-load criterion

This criterion is closely related to the idea of instability and its form is quite simple.

Confining our attention to the tension test, we assume that the constitutive equation of the material can be expressed in the general form

$$\sigma = \sigma(\varepsilon, \dot{\varepsilon}, T, \ldots), \tag{7.4}$$

where σ and ε are the true stress and strain respectively, $\dot{\varepsilon}$ is the strain rate and T is the temperature. The maximum load occurs when

$$dL = \sigma \, dA + A \, d\sigma$$

$$= \sigma \, dA + A \left\{ \left(\frac{\partial\sigma}{\partial\varepsilon}\right)_{\dot{\varepsilon}, T} d\varepsilon + \left(\frac{\partial\sigma}{\partial\dot{\varepsilon}}\right)_{\varepsilon, T} d\dot{\varepsilon} \right.$$

$$\left. + \left(\frac{\partial\sigma}{\partial T}\right)_{\varepsilon, \dot{\varepsilon}} dT + \ldots \right\} = 0. \tag{7.5}$$

If we assume that the effects of strain rate and temperature can be neglected, and taking incompressibility into account, (7.5) becomes

$$\frac{d\sigma}{d\varepsilon} = \sigma.$$

This is the same result as obtained before, and for a material that strain-hardens according to the power law $\sigma = K\varepsilon^n$ the critical strains at diffuse necking and localized necking are n and $2n$ respectively (§§2.1.1 and 2.1.2). A direct comparison between the strains for diffuse and localized necking is shown in Fig. 7.5. If n is decreased then so is the difference between the two critical strains. In particular, when the n-value approaches zero, i.e. the material becomes perfectly plastic, then only local necking occurs.

7.2.3 $\delta\dot{A} = 0$

This principle was introduced by Hart (1967), and the next subsection will be devoted to a discussion of it. The equality means, quite simply, that the rate of reduction of area at a certain location attains a maximum. Compared with the above conditions, this would appear to be intuitively correct, since an area change is involved with localization in tension. So in this sense this condition would seem to be more fundamental from the physical point of view than the others. However, Hart's criterion cannot explain shear bands or characteristics.

Other criteria have been proposed, for example Duncombe (1972) proposed $\delta(\dot{A}/A) = 0$ as a criterion. This implies that the rate of strain at a certain position reaches a maximum (provided that incompressibility is taken into account), which would also seem to be a reasonable criterion. It follows that this situation raises a challenging question: what is a physically reasonable criterion for the onset of localized deformation? Although it will

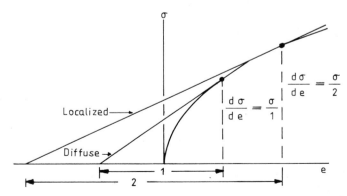

Fig. 7.5   The criteria for localized and diffuse necking in pure tension.

be shown in §7.3 that each of these criteria may be applicable under different conditions, the reality of the situation is that these simple criteria are insufficient and nonunified, and more comprehensive concepts are required. These concepts are bifurcation and instability.

7.2.4 Bifurcation

The discussion of bifurcation is aimed at finding a state at which the solution of a definite boundary-value problem loses its uniqueness. The following discussion concerns a rigid-plastic body and the implications of the expression for bifurcation are highlighted.

The Lévy-von Mises equations for a rigid-plastic solid are given by

$$h\dot{\varepsilon}_{ij} = \begin{cases} n_k \dot{\sigma}_{ki} n_j & (n_k \dot{\sigma}_{ki} \geqslant 0), \\ 0 & (n_k \dot{\sigma}_{ki} \leqslant 0), \end{cases} \tag{7.6}$$

where $\dot{\sigma}_{ij}$ and $\dot{\varepsilon}_{ij}$ are the stress-rate and strain-rate tensors, n_k is the unit outward normal to the yield surface and h is a strain-hardening parameter (see Hill, 1957; and Appendix 7A). Let us suppose that there are two solutions, with subscripts 1 and 2; then it is possible to verify the following inequality at every point in the body:

$$\Delta \dot{\sigma}_{ij} \Delta \dot{\varepsilon}_{ij} \geqslant h(\Delta \dot{\varepsilon}_{ij})^2 \quad \text{provided that } h \geqslant 0, \tag{7.7}$$

where $\Delta \dot{\sigma}_{ij}$ and $\Delta \dot{\varepsilon}_{ij}$ are the differences in the stress and strain rates respectively of the two solutions. This inequality is derived in Appendix 7B.

For an isotropic rigid plastic body with velocity-independent traction rates prescribed on the boundary, the following equality can be derived (Appendix 7C):

$$\int (\Delta \dot{\sigma}_{ij} \Delta \dot{\varepsilon}_{ij}) \, dV = \int \sigma_{kj} \frac{\partial \Delta v_i}{\partial x_k} \frac{\partial \Delta v_j}{\partial x_i} \, dV. \tag{7.8}$$

Combining (7.7) and (7.8) gives as the sufficient condition for uniqueness

$$U = \int \left\{ h(\Delta \dot{\varepsilon}_{ij})^2 - \sigma_{kj} \frac{\partial \Delta v_i}{\partial x_k} \frac{\partial \Delta v_j}{\partial x_i} \right\} dV > 0, \tag{7.9}$$

and the inverse of this inequality provides a lower bound for bifurcation. U is the excess dissipated energy for a pair of virtual modes over the work done by the external force (see Hill, 1958). In view of the energy balance, a sufficient condition imposed on uniqueness is that the dissipated energy should exceed the external work.

Hill's bifurcation criterion has been derived above using some restrictive assumptions, and the mathematical complexity has prevented engineers from using the criterion. Because of this inherent complexity, it may be helpful here to summarize some case studies and comparisons with other criteria without detailing the proofs.

(1) In biaxial tension (plane-stress state) the bifurcation criterion can be satisfied with Swift's maximum-load criterion, namely diffuse necking ($dL_1 = dL_2 = 0$) under all-round dead loading (see Negroni *et al.*, 1968).

(2) Bifurcation usually begins beyond the maximum load (see Needleman, 1972; Hill and Hutchinson, 1975). For an extremely slender column, the bifurcation stress approaches that of the maximum load (see Hutchinson and Miles, 1974).

(3) In a rigid-plastic body, bifurcation of plastic flow may occur before the failure of stability, provided that the strain hardening rate is progressively lowered (see Hill, 1958). For compression of a circular strut, the critical hardening rate for stability is $\frac{1}{5}$ of that for uniqueness.

Other approaches to the evaluation of possible branching points have been made. One approach that seems to have a great potential is that of Stakgold (1971), who proposed that the eigenvalue of a linear problem may signify the branching point of the corresponding nonlinear case.

7.2.5   Instability

An equilibrium state is said to be stable if a disturbance and the resulting additional displacement are both vanishingly small. This is similar to the definition of bifurcation from the energy point of view as long as an additional mode to the equilibrium mode is considered instead of two virtual modes. To do this we use an additional velocity v_i and strain rate $\dot{\varepsilon}_{ij}$ which take the place of the differences Δv_i and $\Delta \dot{\varepsilon}_{ij}$ between virtual modes in the functional U (equation 7.9). Then the criterion obtained is a sufficient condition for stability.

The difference between uniqueness and stability is that the additional strain rate is normal to the yield surface and pointing outwards, but the difference $\Delta \dot{\varepsilon}_{ij}$ of two virtual modes is not necessarily pointing outwards. Thus the condition for bifurcation is wider in scope than that for instability.

Taking the rate of hardening h as a variable parameter and using μh instead of h, where μ is a positive parameter, it is possible to define

$$\left. \begin{array}{ll} H(v_{ij}) = \displaystyle\int h\dot{\varepsilon}_{ij}^2 \, dV, & H(\Delta v_{ij}) = \displaystyle\int h(\Delta \dot{\varepsilon}_{ij})^2 \, dV \\[3mm] \Sigma(v_{ij}) = \displaystyle\int \sigma_{kj} \frac{\partial v_i}{\partial x_k} \frac{\partial v_j}{\partial x_i} \, dV, & \Sigma(\Delta v_{ij}) = \displaystyle\int \sigma_{kj} \frac{\partial \Delta v_i}{\partial x_k} \frac{\partial \Delta v_j}{\partial x_i} \, dV, \\[3mm] \alpha = \max \dfrac{\Sigma(v_{ij})}{H(v_{ij})} \quad \text{in } v_{ij}, & \beta = \max \dfrac{\Sigma(\Delta v_{ij})}{H(\Delta v_{ij})} \quad \text{in } \Delta v_{ij}, \end{array} \right\} \quad (7.10)$$

For a fixed value of h, the criteria for stability and uniqueness according to (7.9) becomes

$$\mu > \alpha, \quad \mu > \beta \tag{7.11}$$

respectively. β is chosen from a wider class of velocity fields that α, whose corresponding strain-rate vectors need to be pointing outwards from the yield surface. It follows that

$$\beta \geqslant \alpha. \tag{7.12}$$

This implies that to maintain uniqueness a higher rate of hardening is required. Because the criteria given by (7.11) are sufficient conditions, the opposite inequalities should be necessary conditions for bifurcation and instability. The condition (7.12) shows that bifurcation may occur at a greater hardening than instability. For a material with a progressively decreasing rate of strain hardening, uniqueness is no longer certain, and bifurcation may occur at an early stage in deformation. This point has already been made in the last section.

Another approach, which is used in fluid dynamics, is based on an essential feature of stability. For a steady velocity field $v_i^0(x_i)$ it is supposed that there is an unsteady infinitesimal disturbance $v_i'(x_i)$. According to this supposition, $v_i' \ll v_i^0$ and the problem reduces to a linear one with unknown v_i. The general solution can be expressed as a supposition of particular solutions, that is the general solution can be written as $e^{i\omega t + k_i x_i}$. Provided that ω contains a negative imaginary part, the term increases with increasing time. This point corresponds to instability. Although the concept is quite straightforward, it is usually rather complex to follow the mathematics.

7.2.6 Band-like localization

Instability and bifurcation do not have direct connections with localization as such. For example, bifurcation is the point beyond which more than one mode of deformation is possible—it does not describe any specific mode.

Hill (1962) and Rice (1977) have shown that a band-like localization can be regarded as a bifurcation and is equivalent to a vanishing propagation velocity of an acceleration wave. This concept can be traced back to Hadamard's idea of a stationary discontinuity. Consider a band as shown in Fig. 7.6 with a normal x_2. Deformation outside the band remains uniform, but changes may occur parallel to x_2. The difference in velocities (inside and outside the band) is

$$\Delta\left(\frac{\partial v_i}{\partial x_j}\right) = g_i(X)n_j, \tag{7.13}$$

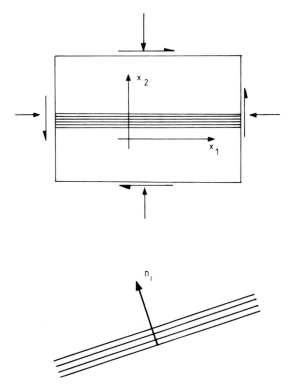

Fig. 7.6 Coordinate system for a band of localization.

where n_j is the normal to the band, Δ denotes the difference between the local field at a point and the uniform field outside the band, $X = n_j x_j$ is the distance along the direction n_j and g is some undetermined function. Combination of the equilibrium equation with the condition for the band requires

$$\Delta \dot{\sigma}_{ij} n_i = 0. \tag{7.14}$$

If the current reference system coincides instantaneously with the current state then

$$\Delta \dot{S}_{ij} n_i = 0, \tag{7.15}$$

where $\dot{S}_{ij}$ is the nominal stress rate.

An acceleration wave front defines jumps in the derivatives of motion such as velocity and strain that remain continuous when crossing the band. Any jump Δf can be differentiated as

$$\frac{d}{dt}(\Delta f) = \Delta\left(\frac{\partial f}{\partial t}\right) + c\Delta\left(\frac{\partial f}{\partial x_j}\right)n_j, \tag{7.16}$$

where c is the front velocity. If $\Delta f = 0$, i.e. if f remains continuous, then (7.16) can be expressed as

$$\Delta\left(\frac{\partial f}{\partial t}\right) = -c\mu, \quad \Delta\left(\frac{\partial f}{\partial x_j}\right) = \mu n_j, \tag{7.17}$$

where μ is arbitrary. Taking $f = v_i$, it follows that

$$\Delta\left(\frac{\partial v_i}{\partial t}\right) = cg_i, \quad \Delta\left(\frac{\partial v_i}{\partial x_j}\right) = g_i n_i, \tag{7.18}$$

provided that $\mu = g_i$. In a similar way, letting $f = S_{ij}$ leads to

$$\Delta\left(\frac{\partial S_{ij}}{\partial x_j}\right) = -\frac{1}{c}\Delta\dot{S}_{ij}n_j. \tag{7.19}$$

Using the equation of motion and substituting (7.18) and (7.19), one obtains

$$\rho c^2 g_i = \Delta\dot{S}_{ji}n_j. \tag{7.20}$$

For nonzero g_i, $c = 0$ provides coincidence between equations (7.15) and (7.20). The implications of the above analysis are: (1) the coincidence of band-like localization and acceleration waves of vanishing speed; and (2) the discussion leads to an eigenvalue problem of an acceleration wave. It seems to be of potential interest to compare this type of approach with the perturbation approach to instability.

At this point it is worthwhile to draw some conclusions from the above discussions.

The maximum load, bifurcation, instability and so forth are all possible criteria for localization, and there are connections between some of the criteria. The discussions of bifurcation and instability that have been reviewed are comprehensive, but the mathematics is complex. Therefore, if possible, simplified criteria are required for engineering use.

Some factors may seriously affect localization; examples of such factors are strain rate, temperature and imperfections. Some of these factors may stabilize uniform deformation, whereas others destabilize deformation. These factors can easily lead to confusion when comparing experimental with theoretical results.

7.3   Necking in uniaxial-tension testing

Here necking of a tensile specimen is treated as a one-dimensional instability. Two fundamental problems are the identification of the onset of instability and the description of the development of the post-instability inhomogeneity and the relationship between this development and the observed ductility.

As mentioned in the previous section there are numerous criteria that are available for the study of necking. Two examples are $\delta \dot{A} = 0$ (Hart, 1967) and $\delta(\dot{A}/A) = \delta(\dot{\varepsilon}) = 0$ (Duncombe, 1972), provided that incompressibility is assumed. Both correspond to kinematical instability, but are not connected with shear banding. Other similar approaches, for example, can be seen in the work of Campbell (1967) and Hutchinson (1979). It is important to be aware of the fact that these criteria presuppose pre-existing defects in the tensile specimen. These defects can be mechanical (hammer-blow defects) or geometrical (machining defects). In the following derivations the constitutive relation inside and outside the necked region is assumed to remain the same. Some basic definitions for uniaxial tension are as follows:

incompressibility

$$ d\varepsilon = \frac{dl}{l} = -\frac{dA}{A} \quad \text{or} \quad \dot{\varepsilon} = \frac{\dot{l}}{l} = -\frac{\dot{A}}{A}, \tag{7.21} $$

where ε denotes the true strain, and l and A are the current gauge length and cross-sectional area respectively;

constitutive equation

$$ \left. \begin{array}{c} \sigma = \sigma(\varepsilon, \dot{\varepsilon}) \quad \text{or} \quad d\sigma = \sigma_\varepsilon d\varepsilon + \sigma_{\dot{\varepsilon}} \, d\dot{\varepsilon}, \\[2mm] \text{where} \quad \sigma_\varepsilon = \left(\frac{\partial \sigma}{\partial \varepsilon}\right)_{\dot{\varepsilon}}, \quad \sigma_{\dot{\varepsilon}} = \left(\frac{\partial \sigma}{\partial \dot{\varepsilon}}\right)_\varepsilon; \end{array} \right\} \tag{7.22} $$

for a constant-load process

$$ dL = 0 \quad \text{or} \quad \frac{d\sigma}{\sigma} = -\frac{dA}{A} = d\varepsilon \quad \text{or} \quad \frac{\dot{\sigma}}{\sigma} = -\frac{\dot{A}}{A}; \tag{7.23} $$

for a process with a constant stretching rate

$$ d\dot{l} = 0 \quad \text{or} \quad \frac{d\dot{\varepsilon}}{d\varepsilon} = -\dot{\varepsilon}. \tag{7.24} $$

Now if we turn our attention to the general forms of the criteria for instability, on substituting (7.21) and (7.22) into $\dot{A}$, we have

$$\delta\dot{A} = -\delta(A\dot{\varepsilon})$$

$$= \frac{A\,\delta\sigma}{\sigma_{\dot{\varepsilon}}}\left(\frac{\sigma_\varepsilon}{\delta\sigma/\delta\varepsilon} + \frac{\sigma_{\dot{\varepsilon}}\dot{\varepsilon}}{\delta\sigma/\delta\varepsilon} - 1\right).$$

Then Hart's criterion provides

$$\sigma_\varepsilon + \sigma_{\dot{\varepsilon}}\dot{\varepsilon} = \frac{\delta\sigma}{\delta\varepsilon}, \tag{7.25}$$

where $\delta\sigma/\delta\varepsilon$ should be understood as a path-dependent differentiation. In a similar way, the criterion $\delta(\dot{A}/A) = \delta(\dot{\varepsilon}) = 0$ is expressed as

$$\sigma_\varepsilon = \frac{\delta\sigma}{\delta\varepsilon}. \tag{7.26}$$

There are two kinematic variables ε and $\dot{\varepsilon}$, and both may be different in the two regions necked and uniformly strained. However, the criterion $\delta(\dot{A}/A) = \delta(\dot{\varepsilon}) = 0$ examines only one factor.

As a further comparison the maximum-load criterion, $dL = 0$ is cited below:

$$\sigma_\varepsilon + \sigma_{\dot{\varepsilon}}\frac{\delta\dot{\varepsilon}}{\delta\varepsilon} = \sigma. \tag{7.27}$$

Obviously, these three criteria depend not only on the constitutive equation but also on the strain path. Only in special cases can some of them be written explicitly as shown in Table 7.1. As path-dependent differentiations exist in

Table 7.1

Instability conditions

Criterion	General form	$L = $ constant	$\dot{l} = $ constant
$\delta L = 0$	$\sigma_\varepsilon + \sigma_{\dot{\varepsilon}}\dfrac{\delta\dot{\varepsilon}}{\delta\varepsilon} = \sigma$	$\gamma + \dfrac{m}{\dot{\varepsilon}}\dfrac{\delta\dot{\varepsilon}}{\delta\varepsilon} = 1$	$\gamma - m = 1$
$\delta\dot{A} = 0$	$\sigma_\varepsilon + \sigma_{\dot{\varepsilon}}\dot{\varepsilon} = \dfrac{\delta\sigma}{\delta\varepsilon}$	$\gamma + m = 1$	$\gamma + m = \dfrac{\delta\sigma}{\sigma\,\delta\varepsilon}$
$\delta\dot{\varepsilon} = 0$	$\sigma_\varepsilon = \dfrac{\delta\sigma}{\delta\varepsilon}$	$\gamma = 1$	$\gamma = \dfrac{\delta\sigma}{\sigma\,\delta\varepsilon}$

$$\gamma = \sigma_\varepsilon/\sigma, \quad m = \sigma_{\dot{\varepsilon}}\dot{\varepsilon}/\sigma$$

the criteria, $\delta L = 0$ is more convenient to use in a constant-pulling-rate test, and $\delta \dot{A} = 0$ is used under dead loading. Therefore the conditions $\gamma - m = 1$ and $\gamma + m = 1$ are appropriate for different testing conditions. It is also possible to examine other types of process, for example the maximum-load criterion for a constant-strain-rate test is simply $\gamma = 1$, which is identical with Duncombe's theory for dead loading.

We shall now turn to the study of the growth of the neck and its relation to fracture strain measurements, following Hart (1967) and Nichols (1980). We confine our attention to the steady growth of a neck under a constant load. Then the growth rate is

$$\left(\frac{d \ln \dot{A}}{d \ln A}\right)_L = -\left(\frac{l \, d\dot{A}}{\dot{l} \, dA}\right)_L, \tag{7.28}$$

after substituting (7.21) into the defined rate. In order to obtain $(l \, d\dot{A}/\dot{l} \, dA)_L$, we first calculate $(d \ln \dot{l}/d \ln l)_L$:

$$\left(\frac{d \ln \dot{l}}{d \ln l}\right)_L = \left(\frac{l \, d\dot{l}}{\dot{l} \, dl}\right)_L = \left\{\frac{d \, (\dot{\varepsilon}l)}{\dot{\varepsilon} \, dl}\right\}_L$$

$$= \left(1 + \frac{l \, d\dot{\varepsilon}}{dl \, \dot{\varepsilon}}\right)_L = \left(1 + \frac{\dfrac{d\sigma}{d\varepsilon} - \sigma_\varepsilon}{\sigma_\varepsilon \dot{\varepsilon}}\right)_L = 1 + \frac{1}{m} - \frac{\gamma}{m}, \tag{7.29}$$

recalling the definitions of γ and m and the definitions given by (7.23).

Steady deformation requires $\gamma = 0$; then

$$\left(\frac{d \ln \dot{l}}{d \ln l}\right)_L = 1 + \frac{1}{m}. \tag{7.30}$$

Another way to derive $(d \ln \dot{l}/d \ln l)_L$ is as follows:

$$\left(\frac{d \ln \dot{l}}{d \ln l}\right)_L = \left(-\frac{A \, d\dot{l}}{\dot{l} \, dA}\right)_L = \left\{\frac{A}{\dot{l} \, dA}\left(\frac{\dot{A}}{A} dl + \frac{l}{A} d\dot{A} - \frac{\dot{A}l}{A^2} dA\right)\right\}_L$$

$$= 2 + \frac{l}{\dot{l}}\left(\frac{d\dot{A}}{dA}\right)_L. \tag{7.31}$$

Comparing (7.30) and (7.31), we obtain

$$-\left(\frac{l \, d\dot{A}}{\dot{l} \, dA}\right)_L = 1 - \frac{1}{m}$$

and

$$\left(\frac{d \ln \dot{A}}{d \ln A}\right)_L = 1 - \frac{1}{m}. \tag{7.32}$$

Integration of (7.31) gives

$$\delta A = \left(\frac{A_0}{A}\right)^{(1/m)-1} A_0, \tag{7.33}$$

provided that $\delta A/\delta A_0 = \dot{A}/A_0$. This indicates that an increasing rate sensitivity m can decrease the neck growth rate.

If it is possible to obtain necking to a point, it follows that $\delta A_f = A_f$, where the subscript f denotes ductile fracture. From (7.33) it follows that

$$\frac{A_f}{A_0} = f^m, \tag{7.34}$$

where $f = \delta A_0/A_0$, which denotes the original defect. Using incompressibility, we obtain

$$e_f = \frac{A_0}{A_f} - 1 = \left(\frac{1}{f}\right)^m - 1, \tag{7.35}$$

where e_f is the engineering fracture strain. Figure 7.7 shows a comparison of the experimental fracture-strain results and strain-rate sensitivity m taken from Nichols (1980) with (7.35) at various levels of initial defects. Other theoretical predictions of e_f that are related to imperfections, rate sensitivity and strain hardening have been made (see e.g. Lin et al., 1981).

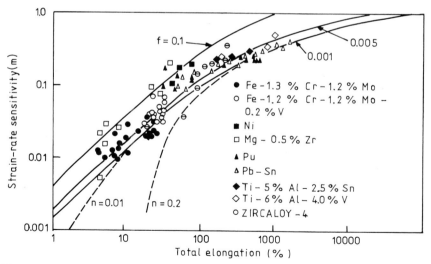

Fig. 7.7 Woodford's correlation and experimentally observed fracture strain and strain-rate sensitivity for various materials and conditions. Solid lines are predicted from (7.35). (After F. A. Nichols, *Acta Metall.* **28**, 663–674 (1980).)

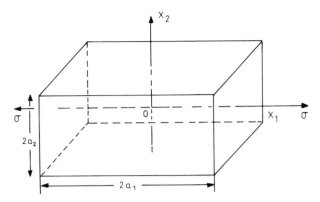

Fig. 7.8 Coordinate system for the plane-strain deformation of a rectangular block.

7.4 Bifurcation in plane-strain tension

To show how the bifurcation analysis is carried out, we consider the tension of a plate of material that is incompressible, incrementally linear and time-independent (see Hill and Hutchinson, 1975). The derivation is not included here because of space considerations, but the important points in the reasoning are given. The interrelations of the physical concepts with the bifurcation analysis are described. For simplicity §§7.4 and 7.5 are concerned only with the plane-strain stress state.

7.4.1 Characteristics, shear banding and bifurcation

The characteristic equation deduced from the equilibrium equation is

$$\{\mu - \tfrac{1}{2}(\sigma_1 - \sigma_2)\}D^4 + 2(2\mu^* - \mu)D^2 + \{\mu + \tfrac{1}{2}(\sigma_1 - \sigma_2)\} = 0, \quad (7.36)$$

where D is the characteristic value, σ_1 and σ_2 are principal stresses, μ and μ^* are the two instantaneous moduli for shear parallel to the axes and at $45°$ to them respectively. The meaning of D can be revealed by discussing an inhomogeneous simple shear normal to $\xi = c_1 x_1 + c_2 x_2 =$ constant, and it is found that $D = c_2/c_1$, where x_1 and x_2 are two axes. Therefore it follows that characteristics are coincident with shear-lines in a similar way to that of slip-lines in a rigid-plastic body. It is possible to classify (7.36) in the following manner: the equation is

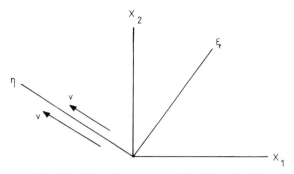

Fig. 7.9 Coordinate system for shear band.

elliptic if there are no real D:

$$2\mu^* > \mu - \{\mu^2 - \tfrac{1}{4}(\sigma_1 - \sigma_2)^2\}^{1/2}; \qquad (7.37a)$$

hyperbolic if there are four real D:

$$2\mu^* < \mu - \{\mu^2 - \tfrac{1}{4}(\sigma_1 - \sigma_2)^2\}^{1/2}; \qquad (7.37b)$$

parabolic if there are two real D:

$$\mu < \tfrac{1}{2}(\sigma_1 - \sigma_2); \qquad (7.37c)$$

for $\mu^* > 0$, $\mu > 0$ and $\sigma_1 > \sigma_2$. In the elliptical domain the shear-band field is excluded.

If we write a variational formula from which the governing equation can be obtained, then we find that

$$U = \int\int u \, \delta x_1 \, \delta x_2 > 0,$$

according to a variational principle (see e.g. Landau and Lifshitz, 1959). This implies that the dissipated energy is greater than the external work. This rules out bifurcation in accordance with the discussion in §7.2.4. In particular, $U > 0$ guarantees uniqueness; that is,

and

$$\left.\begin{array}{c} 0 < \sigma_1 + \sigma_2 < 4\mu^* \\[2mm] \dfrac{\sigma_1^2 + \sigma_2^2}{\sigma_1 + \sigma_2} < 2\mu, \end{array}\right\} \qquad (7.38)$$

which are sufficient conditions for uniqueness.

7.4.2 Bifurcation in uniaxial tension

For simple tension $\sigma_1 = \sigma$, $\sigma_2 = 0$ and the different regimes are

$$\left.\begin{array}{l} 0 < \sigma < 4\mu^*, \\ \sigma < 2\mu, \end{array}\right\} \quad \text{uniqueness;} \tag{7.39}$$

$$\frac{\mu}{2\mu^*} < \frac{\sigma}{4\mu^*} \quad \text{(P) parabolic,}$$

$$1 > \frac{\mu}{2\mu^*} - \left[\left(\frac{\mu}{2\mu^*}\right)^2 - \left(\frac{\sigma}{4\mu^*}\right)^2\right]^{1/2} \quad \text{and} \quad 2\mu > \sigma \quad \text{(E) elliptic,}$$

$$1 < \frac{\mu}{2\mu^*} - \left[\left(\frac{\mu}{2\mu^*}\right)^2 - \left(\frac{\sigma}{4\mu^*}\right)^2\right]^{1/2} \quad \text{and} \quad 2\mu > \sigma \quad \text{(H) hyperbolic.}$$

These relevant regions are shown in Fig. 7.10. For $\sigma < 2\mu$ only elliptic and hyperbolic regimes exist. From conditions (7.38) bifurcation is precluded when $\sigma < 4\mu^*$; this corresponds to conditions prior to the peak load, since the maximum-load condition requires the coincidence of the true stress and the current tangent modulus $4\mu^*$, which owing to $\varepsilon_3 = 0$, gives $\varepsilon_2 = -\varepsilon_1$ in the case we are concerned with. There is a gap between $\sigma = 4\mu^*$ and the elliptic–hyperbolic boundary in the elliptic region. As a shear field is not possible in this region, shear-band localization must occur later than any bifurcation in this gap.

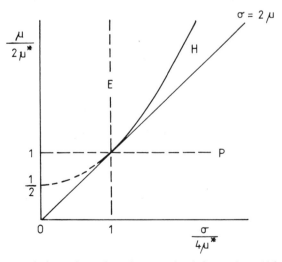

Fig. 7.10 Characteristic regimes for plane-strain deformation. (After R. Hill and J. W. Hutchinson, *J. Mech. Phys. Solids* **23**, 239–264 (1975).)

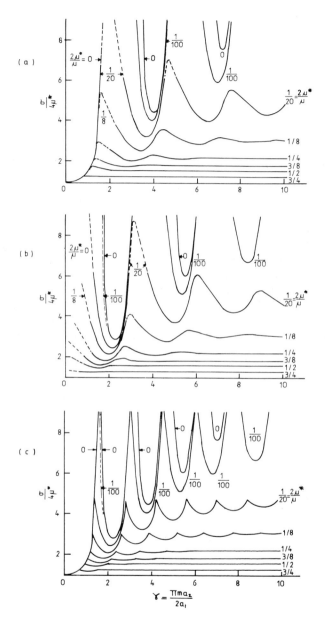

Fig. 7.11 Lower-bound bifurcation stresses for $2\mu^*/\mu < 1$: (a) symmetric modes; (b) antisymmetric modes; (c) symmetric and antisymmetric modes. (After R. Hill and J. W. Hutchinson, *J. Mech. Phys. Solids* **23**, 239–264 (1975).)

The examination of the possibility of a solution, such as the form $\sin(c_2 x_2)\cos(c_1 x_1)$, which is different from the homogeneous solution, provides the detail of bifurcation. Figure 7.11 shows numerical values of the lowest bifurcation stresses for $\mu = 2\mu^*$ and $\gamma = c_1 a_2 = m\pi a_2/2a_1$; solid lines correspond to the elliptic region and dashed lines to the hyperbolic region, there being no values for the parabolic regime. If the symmetrical and anti-symmetrical modes are combined (Fig. 7.11c) then there are only solid lines. Therefore bifurcation can only occur in the elliptic region. When $\mu < 2\mu^*$, $\sigma = 2\mu < 4\mu^*$ bifurcation may be possible below the stress at maximum load at the elliptic/parabolic boundary.

7.5   Shear-band localization and material characterization

For some materials (e.g. structural metals) it is found that diffuse deformation often gives way to highly localized shear bands even for a fairly shallow diffuse neck (see Anand and Spitzig, 1980). Depending on material behaviour and presence and size of imperfections, some shear bands may form catastrophically. These shear bands serve as precursors to rupture because fracture eventually occurs along one of them, so in this section attention is concentrated only on shear bands.

In the later discussions of localization three hardening theories will be introduced. These are shown in Fig. 7.12. If during the course of plastic flow the yield surface expands uniformly in stress space then this is referred to as isotropic hardening (Fig. 7.12a). When the loading surface translates in stress space similar to a rigid body, i.e. retaining its shape, size and orientation, this is kinematic hardening. Corner theories are based on the concept that the loading surface sustains a distortion during the course of plastic flow, a corner developing at the loading point (Fig. 7.12c).

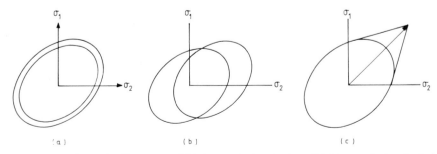

Fig. 7.12   Schematic diagrams of (a) isotropic hardening, (b) kinematic hardening and (c) corner theory.

7.5.1   Inclination of shear bands

Theoretically it has recently been realized that shear-band localization is very sensitive to material characterization. For example, the onset of shear bands is influenced by deviations from a smooth yield surface (e.g. development of corners), dilation and imperfections.

A prerequisite for shear-band localization is a loss of ellipticity. From (7.39) and Fig. 7.10, the condition for this is

or
$$\left.\begin{array}{ll} \sigma > 2\mu & E \to P \\ \sigma^2 > 16\mu^*(\mu - \mu^*) & E \to H. \end{array}\right\} \qquad (7.40)$$

Provided that $\mu > 4\mu^*$, ellipticity is lost owing to (7.40). The angle of inclination of the shear band to the pulling axis, ϕ, is given (from 7.36) by

$$D^2 = \left(\frac{c_2}{c_1}\right) = \cot^2 \phi$$

$$= 1 + \frac{\sigma - 4\mu^* \pm \{(\sigma - 4\mu^*)^2 - 8\mu^*(2\mu - \sigma)\}^{1/2}}{2\mu - \sigma} \qquad (7.41)$$

Because $\mu > \mu^*$, and from (7.40) and (7.41), it follows that

$$\cot^2 \phi > 0 \quad \text{or} \quad \phi < \tfrac{1}{4}\pi$$

for plane-strain tension $\sigma > 0$.

7.5.2   The effect of material on shear-band localization

The critical condition for possible shear-band formation is

$$\sigma_c^2 = 16\mu^*(\mu - \mu^*). \qquad (7.42)$$

This occurs when

$$\tan^2 \phi = \frac{2\mu - 4\mu^*}{2\mu + \sigma}. \qquad (7.43)$$

Rice (1977) proposed that only these constitutive parameters, which are a measure of stiffness, control the occurrence of localization. Since shear-band localization imposes deformation of increments orthogonal to the plastic flow, it induces an elastic response in a classical elastic-plastic body with a smooth yield surface and μ is necessarily elastic. However, $4\mu^*$ is a plastic modulus. Then it follows that usually $4\mu^* \ll \mu$, and from (7.42),

$$\sigma_c \sim 4(\mu\mu^*)^{1/2}. \qquad (7.44)$$

Equation (7.44) shows that the critical stress for a classical elastic-plastic solid with a smooth yield surface is not low when μ is large.

If a corner develops on the yield surface during plastic flow, the response of the material to the change in the shear modes should not be elastic; then μ can be less than the elastic shear modulus and σ_c can be much lower. In the same way, the critical strain also decreases.

Alternatively, if the deformation theory of plasticity or nonlinear elastic theory is used then a low critical strain for shear-band formation is obtained. For example, if power-law hardening is assumed with a small n-value then deformation theory gives $\varepsilon_c \sim n^{1/2}$, which is quite small (see Hutchinson and Tvergaard, 1981).

7.5.3 Comparison of experiments with theoretical studies

Anand and Spitzig (1980) carried out an experimental study of the emergence of shear bands in a maraging steel in tension and compression tests. They used microscopy to examine polished and etched cross-sections of specimens so that they could detect the first appearance of shear bands. They detected the critical strains and angles of inclination of shear bands given in Table 7.2.

Table 7.2

		Observed	Predicted	
			Deformation theory	Flow theory
Tension	ε_c	$+0.034$	$+0.085$	$+0.184$
	ϕ	$\pm(38 \pm 2)°$	$\pm 42.55°$	$\pm 45°$
Compression	ε_c	-0.034	-0.085	-0.184
	ϕ	$\pm(55 \pm 2)°$	$\pm 47.45°$	$\pm 45°$

In order to compare experimental observations with theoretical predictions, two simple models are used: a flow theory and a deformation theory. It is clear that deformation theory provides closer values than flow theory. From (7.42) the value of $45°$ may be due either to the high elastic value of μ in the flow theory or a small critical strain in the deformation theory (see Hutchinson and Tvergaard, 1981).

7.5.4 Effect of imperfections on shear-band localization

Yield surfaces with corners and deformation theory are not the only possibilities that can lower the critical strain for the onset of shear-banding. The presence of imperfections and dilatation due to voids are other possibilities.

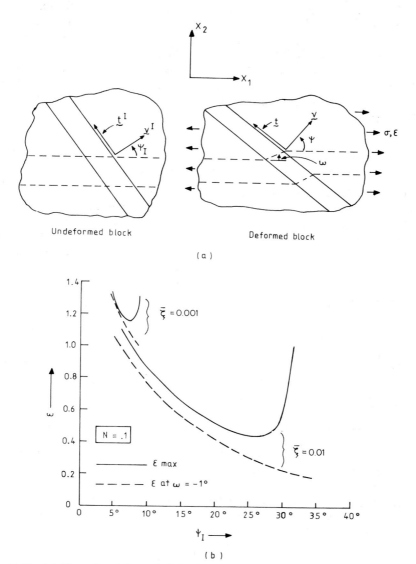

Fig. 7.13 (a) Shear band in an infinite block of material. (b) Localization strain ε versus initial band orientation, ψ_1 for a kinematically hardening solid with $\sigma_y/E = 0.005$ and index $N = 0.1$. (After J. W. Hutchinson and V. Tvergaard, *Int. J. Solids and Structures* **17**, 451–470 (1981).)

Hutchinson and Tvergaard (1981) analysed the role of imperfections in shear-band localization in a kinematically hardening solid with a smooth yield surface. The imperfection has an angle ψ_1 as shown in Fig. 7.13, and $\bar{\xi} = (\sigma_y^0 - \sigma_y^b)/\sigma_y^0$ where σ_y^0 and σ_y^b denote the yield stresses within and outside the band respectively. The analysis provides the dependence of the localization strain on the initial angle of inclination ψ_1 and inhomogeneity $\bar{\xi}$ (Fig. 7.13a). The dashed line (Fig. 7.13b) corresponds to the strain when the rotation in the band attains $\omega = -1°$. Shear localization occurs just beyond this small rotation. The greater the inhomogeneity $\bar{\xi}$, the smaller the critical strain ε_c. The important point here is that the critical strain has realistic values. However, if ψ_1 is too large then localization will not occur.

The effect of voids on shear-band localization has been studied by, for example, Yamamoto (1978) and Rudnicki and Rice (1975). Later, Tvergaard (1981) used Gurson's continuum model (see §5.2.3) in a study of the onset of shear localization. In addition, Saje et al. (1982) carried out a similar study with Gurson's model and a model of void nucleation, which is similar to formula (5.35). All these studies reveal that the onset of localization is highly void-sensitive. It may well be that the nucleation of voids is more important than void growth in shear-band localization. This is so because a burst of void nucleation would reduce the tangent moduli and therefore make the material more susceptible to shear band localization (see Hutchinson, 1981). The development of shear bands is another very interesting and challenging problem. Band development is also a function of the material. For example, using deformation theory the shear band develops catastrophically. On the other hand, shear-band development is not like this for a material with a corner on the yield surface. In the presence of a band-like imperfection, the band eventually localizes in a solid that hardens kinematically in a similar way to a material with a yield-surface corner. However, if the material hardens isotropically then the band does not develop (see Hutchinson, 1981).

Tvergaard et al. (1981) numerically analysed the effect of thickness of inhomogeneities on flow localization for a nonlinear solid and a solid with a yield-surface corner in plane-strain tension. Results of these calculations for these two material models are shown in Fig. 7.14, which shows deformed meshes and contours of constant maximum strain.

The susceptibilities of shear-band localization to destabilizing effects of yield-surface shape, mode of hardening, dilatation, void nucleation and growth, imperfections, and non-normality of strain-rate vector etc. are not yet well understood. However, it can be concluded that shear-band localization is very sensitive to material characterization. The existence of crack-like shear bands rather than instantaneous shear-band fields, which are initiated in some cases, challenges the assumptions of the conventional bifurcation analysis. Also, the concept of stationary acceleration waves introduced in §7.2.6 requires more work for the full development of the theory.

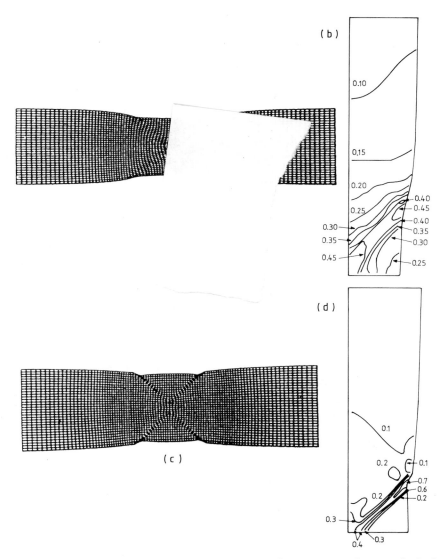

Fig. 7.14 The deformed mesh (a,c) and contours of constant maximum principal logarithmic strain in the deformed configuration (b,d), with the initial thickness inhomogeneity specified by $\xi_1 = 0.375 \times 10^{-2}$, $\xi_2 = 0.15 \times 10^{-2}$. (a,b) are for a J_2 corner-theory computation. (c,d) are for a nonlinear elastic computation. (After V. Tvergaard *et al.*, *J. Mech. Phys. Solids* **29**, 115–142 (1981).)

7.6   Thermal rate-dependent shear bands

Here we shall describe a special kind of shear-band localization that includes the effects of strain rate and heat resulting from the plastic work. Some practical examples of this mode of localization have been described in Chapter 6. The maximum current flow stress was adopted there, and, despite the restricted assumptions made, the concept has been shown to be valuable. The mathematics involving this effect can easily become extremely complicated in the more general cases. Therefore here we shall confine our attention to simple shear. Bai (1980, 1982) took into account the effect of strain rate, the heat produced by plastic work and heat conduction in his development of an approximate model of simple shear. The model can be described by

$$\left. \begin{aligned} \rho \frac{\partial^2 \gamma}{\partial t^2} &= \frac{\partial^2 \tau}{\partial y^2}, \\ K\tau \frac{\partial \gamma}{\partial t} &= \rho c_v \frac{\partial T}{\partial t} - \lambda \frac{\partial^2 T}{\partial y^2}, \end{aligned} \right\} \tag{7.45}$$

where ρ is the density, c_v is the specific heat, λ is the thermal conductivity, T is the temperature, and τ and γ are the shear stress and strain respectively. K is taken to be a constant ($K \approx 0.9$–0.95) denoting the portion of plastic work converted into heat. The y-axis is parallel to the normal to the band.

The problem is to discover the conditions under which a quasi-steady uniform-shear deformation gives way to localization. It is assumed that localization can be considered as an instability of the solution. Following the perturbation approach mentioned in §7.2.5 and considering disturbances of the type $\gamma_a\, e^{\alpha t + iky}$, it is possible to obtain the following homogeneous equations from (7.45):

$$\left. \begin{aligned} \rho \alpha^2 \gamma_a + k^2 . \tau_a &= 0, \\ K\tau_0 \alpha \gamma_a + K\dot{\gamma}_0 \tau_a - (\rho c_v \alpha + \lambda k^2) T_a &= 0, \end{aligned} \right\} \tag{7.46}$$

where the subscripts 0 and a denote the basic state and the amplitude of the disturbance respectively. The spectral equation of α is given when the determinant of the homogeneous equations is equated to zero. Taking the constitutive equation as

$$\tau = \tau(\gamma, \dot{\gamma}, T), \tag{7.47}$$

and denoting

$$Q = \left(\frac{\partial \tau}{\partial \gamma} \right)_0, \quad R = \left(\frac{\partial \tau}{\partial \dot{\gamma}} \right)_0, \quad P = -\left(\frac{\partial \tau}{\partial T} \right)_0,$$

the following spectral equation is obtained:

$$\rho^2 c_v \alpha^3 + \{KP\dot{\gamma}_0 + (\lambda + c_v R)k^2\}\alpha^2$$
$$+ (\lambda R k^2 + \rho c_v Q - K\tau_0 P)k^2\alpha + \lambda Q k^4 = 0. \qquad (7.48)$$

A positive α implies an increasing magnitude of the disturbance with time, i.e. instability. It is important to appreciate that α depends upon physical properties of the material, such as ρ, c_v and λ, as well as the constitutive relation, the strain state, P, Q, R and the wavenumber k. For all possible disturbances $(0 < k < \infty)$, as long as

$$\frac{K\tau_0 P}{\rho c_v Q} > 1 + 2\left(\frac{K\lambda P\dot{\gamma}_0}{\rho c_v^2 Q}\right)^{1/2}, \qquad (7.49)$$

then $\alpha > 0$ provided that $Q > 0$. For the special case of adiabatic deformation $\lambda = 0$, instability occurs when

$$K\tau_0 P = \rho c_v Q, \qquad (7.50)$$

which coincides with the answer obtained in Chapter 6 with $d\tau = 0$. However, from (7.49) it can be seen that both the thermal conductivity λ and the current strain rate $\dot{\gamma}_0$ are stabilizing factors. This seems physically reasonable.

Numerous observations (see e.g. Rogers, 1979; Olson et al., 1981) show that the width of the so-called thermoplastic shear band in metals is normally of the order of tens of micrometres, within which the strain rate can be as high as $10^7 \, \text{s}^{-1}$ and traces of melting are sometimes observed (see §6.3). At these very high strain rates the material is very strain-rate-sensitive and it is almost possible to disregard strain (see Campbell, 1973). Therefore deformation may reach a steady regime provided that the constitutive relation includes viscosity, such as $\dot{\gamma} = g(\tau, T)$ (see Bai et al., 1984). At this stage a band-like solution is expressed as

$$y = \int_T^{T_c}\left[\int_\xi^{T_c} \frac{2K\tau_c g(\tau_c, \xi')}{\lambda} \, d\xi'\right]^{-1/2} d\xi, \qquad (7.51)$$

where T_c denotes the temperature at the centre of the band and τ_c is the stress within the band. For simplicity, the width of a band can be estimated as

$$\delta^2 = \frac{\lambda T^*}{\tau^* \dot{\gamma}^*}, \qquad (7.52)$$

where the asterisk denotes characteristic quantities within the band. The significance of (7.52) is obvious—increasing thermal conductivity increases the band width while the opposite is so when the plastic work rate increases.

Eventually a steady-state thermoplastic shear band results when these two effects balance.

A similar perturbation approach was proposed by Clifton (1980), who considered a deformation mode similar to the above, but used a different constitutive relation that included the activation form of the shear strain and took thermal expansion into account. After appropriate manipulations and assuming that yielding is pressure-insensitive, the criterion for instability is

$$m\dot{\gamma}_0 \left\{ \frac{1}{\tau_r} \left(\frac{\partial \tau_r}{\partial \gamma_p} \right) + \frac{K}{\rho c_v} \left(\frac{\partial \tau}{\partial T} \right) \right\} + \frac{\lambda \xi^2}{\rho c_v} < 0, \tag{7.53}$$

where $m = \partial \ln \dot{\gamma}_p / \partial \ln \tau$, τ_r represents the resistance of the material to plastic deformation and ξ^{-1} is the wavelength of the inhomogeneity. The relative magnitudes of the first two terms are essential in the stability discussion. In the case of a negligible thermal conductivity λ or a high strain rate, the condition becomes the same as (7.50) provided that we take τ_r as the current flow stress. This feature shows the physical importance of a balance between hardening and thermal softening. The terms including λ in (7.49) and (7.53) are different because of the different assumptions about material behaviour.

Although simple shear is a very simple mode of deformation, it is still not possible to fully understand thermoplastic shear instability except for the essential balance between hardening and thermal softening. Also, there is still too little experimental data to compare reliably with the theory. Further, the linearized perturbation technique can only deal with the occurrence of a simultaneous instability. Clearly a knowledge of the evolution of shear bands is required so that it is possible to compare experimental observations with theoretical models.

Appendix 7A: The meaning of h

The Lévy–von Mises constitutive relations for a rigid plastic body using Hill's (1957) notation become

$$h\dot{\varepsilon}_{ij} = \begin{cases} (n_k \dot{\sigma}_{ki})n_j & (n_k \dot{\sigma}_{ki} > 0), \\ 0 & (n_k \dot{\sigma}_{ki} < 0). \end{cases} \tag{A1a} \tag{A1b}$$

Assuming that the material is isotropic and taking $n_i n_i = 1$, (A1) reduces to

$$h\dot{\varepsilon}_{ij} = \begin{cases} \dot{\sigma}_{ij} & (n_k \dot{\sigma}_{ki} > 0), \\ 0 & (n_k \dot{\sigma}_{ki} < 0). \end{cases} \tag{A2a} \tag{A2b}$$

The Lévy–von Mises equations in terms of the plastic potential are

$$\dot{\varepsilon}_{ij} = \dot{\lambda} \frac{\partial f}{\partial \sigma_{ij}}. \tag{A3}$$

Substituting into (A2), for the loading one obtains

$$h\dot{\lambda} = \dot{\sigma}_{kl} \frac{\partial f}{\partial \sigma_{kl}} \left/ \frac{\partial f}{\partial \sigma_{ij}} \frac{\partial f}{\partial \sigma_{ij}} \right. \tag{A4}$$

and

$$h\dot{\varepsilon}_{ij} = \frac{\dfrac{\partial f}{\partial \sigma_{kl}} \dot{\sigma}_{kl}}{\dfrac{\partial f}{\partial \sigma_{mn}} \dfrac{\partial f}{\partial \sigma_{mn}}} \frac{\partial f}{\partial \sigma_{ij}}. \tag{A5}$$

In order to give h and $\dot{\lambda}$ definite forms, it is necessary to define the potential f. An obvious choice for f is the von Mises yield condition

$$f = \tfrac{1}{2}\sigma'_{ij}\sigma'_{ij}. \tag{A6}$$

Hence

$$h\dot{\varepsilon}_{ij} = \frac{\dot{f}}{\sigma'_{mn}\sigma'_{mn}} \frac{\partial f}{\partial \sigma_{ij}}, \tag{A7}$$

and (A5) reduces to

$$h\dot{\varepsilon}_{ij} = \frac{\dot{f}}{2f} \frac{\partial f}{\partial \sigma_{ij}}. \tag{A8}$$

After introducing the plastic work rate $\dot{W}_p$, (A8) becomes a scalar:

$$h\dot{W}_p = h\sigma_{ij}\dot{\varepsilon}_{ij} = \dot{f}\sigma_{ij}\frac{\partial f}{\partial \sigma_{ij}} \left/ 2f, \right. \tag{A9}$$

so that

$$h\dot{W}_p = \dot{f}\frac{\dot{W}_p}{\dot{\lambda}} \left/ 2f. \right. \tag{A10}$$

In the Lévy–von Mises equations using the von Mises yield condition as a plastic potential, $\dot{\lambda}$ is usually expressed in terms of the effective strain rate $\dot{\bar{\varepsilon}}$ and stress $\bar{\sigma}$:

$$\dot{\bar{\varepsilon}} = (\tfrac{2}{3}\dot{\varepsilon}_{ij}\dot{\varepsilon}_{ij})^{1/2}, \tag{A11}$$

$$\bar{\sigma} = (\tfrac{3}{2}\sigma'_{ij}\sigma'_{ij})^{1/2}; \tag{A12}$$

therefore

$$\tfrac{3}{2}\dot{\bar{\varepsilon}}^2 = \tfrac{2}{3}\bar{\sigma}^2\dot{\lambda}^2. \tag{A13}$$

Bearing in mind that $f = \frac{1}{3}\bar{\sigma}^2$, one obtains

$$\left. \begin{aligned} h &= \frac{2}{3}\frac{\dot{\bar{\sigma}}}{\dot{\bar{\varepsilon}}} \\[2mm] h &= \frac{2}{3}\frac{d\bar{\sigma}}{d\bar{\varepsilon}}. \end{aligned} \right\} \tag{A14}$$

or

Appendix 7B: Derivation of $\Delta\dot{\sigma}_{ij}\,\Delta\dot{\varepsilon}_{ij} \geqslant h(\Delta\dot{\varepsilon}_{ij})^2$

When $h > 0$ we have the following.

Case 1

$$\dot{\varepsilon}_{ij}^{(1)} \neq 0, \quad \dot{\varepsilon}_{ij}^{(2)} = 0 \quad \text{from (A2a)};$$

$$(\dot{\sigma}^{(2)} - \dot{\sigma}^{(1)})(\dot{\varepsilon}^{(2)} - \dot{\varepsilon}^{(1)}) = h(\dot{\varepsilon}^{(2)} - \dot{\varepsilon}^{(1)})^2. \tag{B1}$$

Case 2

$$\dot{\varepsilon}^{(1)} \neq 0, \quad \dot{\varepsilon}^{(2)} = 0 \quad \text{from (A2a)};$$

$$(\dot{\sigma}^{(2)} - \dot{\sigma}^{(1)})(\dot{\varepsilon}^{(2)} - \dot{\varepsilon}^{(1)})$$

$$= -(\dot{\sigma}^{(2)} - \dot{\sigma}^{(1)})\dot{\varepsilon}^{(1)} = \frac{(\dot{\sigma}^{(1)} - \dot{\sigma}^{(2)})\dot{\sigma}^{(1)}}{h}$$

$$= \frac{\{\dot{\sigma}^{(1)}\}^2}{h} - \frac{\dot{\sigma}^{(2)}\dot{\sigma}^{(1)}}{h} = h\{\dot{\varepsilon}^{(1)}\}^2 - \frac{\dot{\sigma}^{(2)}\dot{\sigma}^{(1)}}{h}$$

$$\geqslant h\{\dot{\varepsilon}^{(1)}\}^2, \tag{B2}$$

according to (A2*b*).

Case 3

$$\dot{\varepsilon}^{(1)} = 0, \quad \dot{\varepsilon}^{(2)} \neq 0, \tag{B3}$$

which is similar to Case 2.

Case 4

$$\dot{\varepsilon}^{(1)} = \dot{\varepsilon}^{(2)} = 0;$$

then

$$(\dot{\sigma}^{(2)} - \dot{\sigma}^{(1)})(\dot{\varepsilon}^{(2)} - \dot{\varepsilon}^{(1)}) = 0 = h(\dot{\varepsilon}^{(2)} - \dot{\varepsilon}^{(1)})^2. \tag{B4}$$

Appendix 7C: Derivation of $\int (\Delta \dot{\sigma}_{ij} \, \Delta \dot{\varepsilon}_{ij}) \, dV = \int \sigma_{kj} \dfrac{\partial \Delta v_i}{\partial x_k} \dfrac{\partial \Delta v_j}{\partial x_i} \, dV$

The formulation of the problem with velocity-independent traction rates prescribed on the boundary is as follows:

$$\left. \begin{aligned} \frac{\partial \dot{\sigma}_{ij}^0}{\partial x_i} &= 0 \quad \text{in the body,} \\[2mm] T_j^0 &= \begin{cases} n_i \sigma_{ij}^0 & \text{on surface } A_\mathrm{T}^0, \\ v_j & \text{on surface } A_v^0, \end{cases} \end{aligned} \right\} \tag{C1}$$

where superscript 0 denotes nominal quantities. In accordance with incompressibility and the assumption that the current reference axes should be coincident with the initial axes, the true stress is

$$\sigma_{ij} = \sigma_{ij}^0, \tag{C2}$$

and the rates of nominal and true stresses are connected by

$$\dot{\sigma}_{ij}^0 = \dot{\sigma}_{ij} - \sigma_{kj} \frac{\partial v_i}{\partial x_k}. \tag{C3}$$

Assuming that there are two virtual rate field modes satisfying the boundary-value problem, one can write

$$\frac{\partial \Delta \dot{\sigma}_{ij}^0}{\partial x_i} = 0 \quad \text{in } V, \tag{C4a}$$

$$\Delta \dot{T}_j^0 = n_i^0 \, \Delta \dot{\sigma}_{ij}^0 = 0 \quad \text{on } A_\mathrm{T}^0, \tag{C4b}$$

$$\Delta \dot{\sigma}_{ij}^0 = \Delta \dot{\sigma}_{ij} - \sigma_{kj} \frac{\partial \Delta v_i}{\partial x_k}. \tag{C5}$$

From the boundary conditions $\Delta \dot{T}_j^0 = 0$ on A_T^0 and $\Delta v_j = 0$ on A_v^0, the integral $\int \Delta \dot{T}_j^0 \, \Delta v_j \, dA^0$ is identically zero. By Gauss's theorem and the equilibrium equation, this gives

$$\int \frac{\partial}{\partial x_k} (\Delta \dot{\sigma}_{kj}^0 \, \Delta v_j) \, dV = \int \Delta \dot{\sigma}_{kj}^0 \frac{\partial \Delta v_j}{\partial x_k} \, dV$$

$$= 0. \tag{C6}$$

Substituting (C5) into (C6) and recalling the assumption of symmetry made in (A1) gives the required equality.

8

Metalworking Processes and Workability

In the previous chapters we have described in some depth fundamental aspects of ductile fracture. Environmental and microstructural variables that influence fracture have been discussed, as well as theoretical approaches to ductile fracture.

In metal-forming processes large plastic deformations are commonly achieved. In these practical plastic deformation processes all of the following can vary: strain path, strain rate, flow stress, stress state and temperature. Thus it may well be that the theoretical approaches described earlier will be difficult, if not impossible, to apply to these processes with reliability. However, in the following chapters the ductility exhibited by metals in forming processes, that is the workability of the metal, is discussed in relation to the theories described above.

8.1 Metalworking processes

Metalworking processes achieve shape changes by either plastic deformation of the workpiece or a combination of plastic deformation and cracking. Examples of processes in the first category (generally termed metal-forming) are extrusion, drawing, rolling and forging, and examples of processes that involve cracking are blanking, cropping, shearing and cutting. In shaping processes such as forging the onset of ductile fracture is a major limitation, and consequently fracture must be avoided if these processes are to be used successfully. On the other hand, for processes in the second category, which involve separation of the workpiece, a material in which cracks can nucleate and grow quickly will separate easily.

Classification of processes

Metalworking processes can be classified according to the forces applied to the workpiece, after Sachs (1954). Using this classification there are five basic categories of process:

(1) direct compression processes;
(2) indirect compression processes;
(3) tension processes;
(4) bending processes;
(5) shearing processes.

Examples from these categories are shown in Fig. 8.1. Forging and rolling are examples of direct-compression processes. In the second category, which includes wire drawing, the applied forces are tensile, but large indirect compressive forces are generated by the dies.

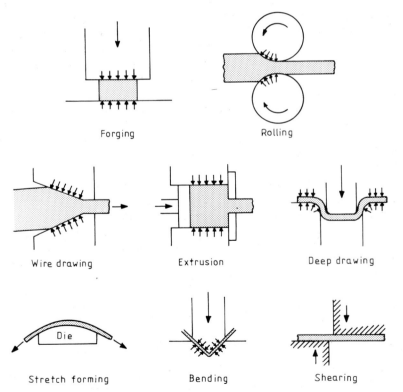

Fig. 8.1   Some metal-forming processes. (After G. E. Dieter, *Mechanical Metallurgy*, 2nd edn. McGraw-Hill, New York, 1976.)

A similar but more informative classification of metal-forming processes has been described by Johnson (1971), who paid special attention to the identification of shear stresses (or stress deviators) and the hydrostatic stress acting on an element in forming processes. As described earlier, the shear stresses are responsible for the shape change in the workpiece, while the hydrostatic stress influences material ductility or formability. We have seen how a hydrostatic pressure enhances material ductility by suppression of void nucleation and growth, and conversely a tensile hydrostatic stress promotes material fracture.

Yet another method of classifying working processes is into the two broad categories of cold and hot working. Plastic deformation in cold-working processes causes grain distortion as it is carried out below the recrystallization temperature of the work material. On the other hand, hot working is carried out at temperatures such that recrystallization occurs in the workpiece. As listed by Dieter (1976) and Harris (1983), cold-worked metals normally: (a) have clean surfaces; (b) have good tolerances; (c) have high hardness and strength as well as low ductility; (d) the mechanical properties of the work material are often anisotropic; and as we have noted above (e) the grain structure is distorted. Hot worked metals normally: (a) have an oxidized surface; (b) the tolerances are not as good as for cold-worked products; (c) the products are ductile and have a low strength; (d) the mechanical properties tend to be more isotropic than in the cold-worked product because of recrystallization.

As an example of the many different possible processing routes, we shall consider the manufacture of a cup.

Figure 8.2 show schematically the various processing routes which can be used to obtain a cup from a melt. Slab can be continuously cast from the melt, hot rolled and then may be either cold rolled and sheared or cold rolled, blanked and deep drawn. An alternative processing path is to cast an ingot and bloom roll it to form a billet, which is then hot rolled, cropped, upset and cold extruded. Which one of these routes will be used will depend on numerous factors, not least of which will be equipment availability and cost as well as workability of the material at each stage in the process. It may be that the workability is limited in one particular operation (e.g. upsetting), which precludes processing routes which include this step.

8.2 Typical microstructures related to metal processing

Sachs (1954) attempted to classify the microstructures commonly found in alloys that consist of two different phases. The microstructure shown in Fig. 8.3(b) is often referred to as a dispersed structure (as opposed to an aggregated structure shown in Fig. 8.3(a)).

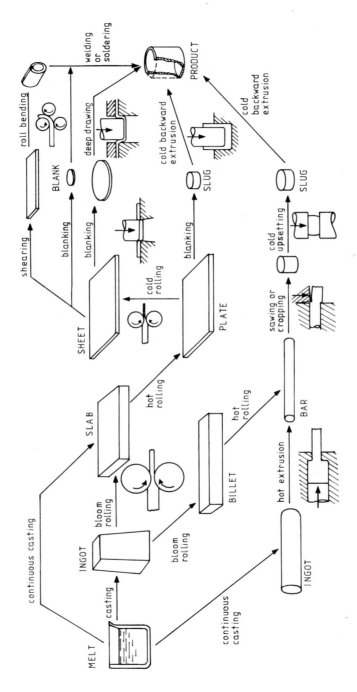

Fig. 8.2 Examples of processing routes from melt to cup. (By courtesy of H. Kudo.)

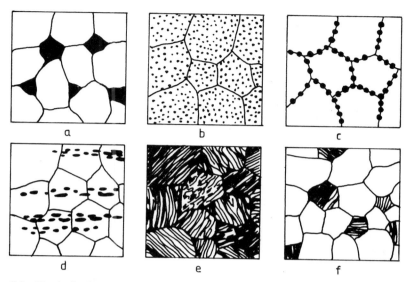

Fig. 8.3 Typical microstructures of metals containing two difference phases: (a) basic aggregated heterogeneous structure; (b) the second phase is embedded in a matrix of the other; (c) the second phase is spread along the grain boundaries of the other; (d) the second phases are in the form of stringers spread in the direction of prior working; (e) typical eutectic or eutectoid structure consisting of grains of alternate lamellae of the two phases; (f) a hypo- or hypereutectic/oid structure having excess of one phase. (After G. Sachs, *Fundamentals of the Working of Metals*. Pergamon Press, Oxford, 1954.)

The eutectoid transformation in carbon steels may produce many different microstructures, depending on the thermomechanical history of the steel. The eutectoid transformation can affect the cold workability of steels, but the hot workability is usually not affected. Whether a particular steel will perform well in a cold-forming operation will depend upon the arrangement of the two constituents ferrite and cementite. A number of possible structural configurations are shown in Fig. 8.4. Spheroidized structures such as that shown in Fig. 8.4(c) are soft and ductile. Cold-drawn wire (see Fig. 8.4d) can have a strength in excess of heat-treated steels.

A fibred structure as shown in Fig. 8.4(a) may be produced by plastic working within a temperature range where its structure is heterogeneous, in which case there is a danger that the grains of the harder constituent will break up and spread out to form a fibred or banded structure. This can be overcome if the alloy is annealed in the austenite phase field. In this case the banded structure disappears and a structure similar to that shown in Fig. 8.4(b) is obtained.

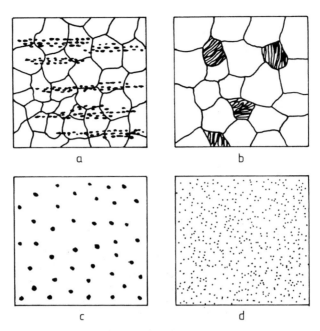

Fig. 8.4   Typical microstructures of carbon steels, the dark particles are cementite: (a) after process annealing; (b) after normalizing; (c) after spheroidizing; and (d) after patenting. (After G. Sachs, *Fundamentals of the Working of Metals*. Pergamon Press, Oxford, 1954.)

In a detailed review of the microstructure of alloys, Hornbogen (1984) showed degrees of mechanical anisotropy in various microstructures (see Fig. 8.5). Clearly the anisotropy of the laminated and fibred structure can be marked, and this in itself may profoundly affect material workability. The equiaxed duplex and skeleton structures will both be fairly isotropic mechanically. More complex microstructures are of course possible; for example: (1) those exhibiting short- and long-range order; (2) arrays of dislocations and particles; (3) mixtures of dispersions plus dislocations and particles; (4) partial or complete transformation to a nonequilibrium martensitic structure. Figure 8.6 shows various mathematical transformations of a two-phase structure. An affine transformation is shown in Fig. 8.6(b), and a more complex topological transformation is shown in Fig. 8.6(c). When the volume fraction of the second phase β becomes larger than that of the first phase α, then β forms the matrix instead of α; compare Figs. 8.6(a) and (d).

It is important to attempt to link workability indices with microstructural features, such as the spacing, size and shape of inclusions and/or second-

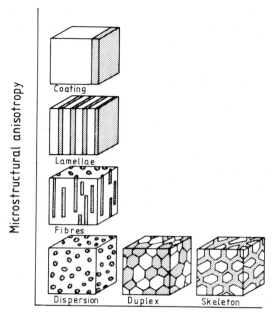

Fig. 8.5 Types of two-phase microstructure. (After E. Hornbogen, *Acta Metall.* **32**, 615–627 (1984).)

phase particles. Linking ductility with microstructure has been discussed in Chapter 4; a similar link between workability and microstructure is more complex because of additional considerations—particularly the process variables.

8.3 Workability

Workability is a very difficult term to define, as it depends on the ductility of the material being worked and the stress, strain, strain rate and temperature distributions in the workpiece. In turn, these factors are dependent on process variables such as geometry of tooling and workpiece as well as lubrication. In many cases the workability limit can be defined as the limiting plastic deformation beyond which cracking or fracture occurs in the workpiece. Examples of such fractures are centre bursts or chevron cracking in extrusion, surface and body cracking in upset forging, and cup fracture in deep drawing.

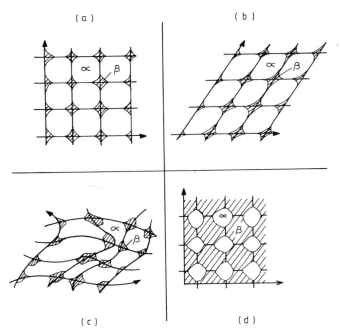

Fig. 8.6 Mathematical transformations of two-phase microstructures: (a) original structure; (b) affine transformation; (c) topological transformation; (d) matrix phase change. (After E. Hornbogen, *Acta Metall.* **32**, 615–627 (1984).)

Other limitations on the degree of plastic deformation that may be set are the onset of orange peeling and the beginning of grain coarsening in hot working. A further important property of pressed metal sheets is their fittability, that is the degree to which the components are dimensionally accurate (Ishigaki, 1978). After pressing, how much elastic springback is there in the components? Thus fittability is another factor that is of importance in the overall workability of a material. Other factors that may be important in the context of workability are surface finish and degree of die filling.

Some of the typical defects found in drawn components are shown in Fig. 8.7. Some of these defects are also workability limits; for example, as well as the obvious limitations set by fracturing, others such as wall wrinkling, burnishing and the appearance of stretcher strains may also be important.

Internal damage in the work material may set a significant limit to plastic deformation; this is particularly so for load-bearing components made of alloys containing second-phase particles. An extensive study of internal structural damage in processing has been carried out by Rogers (1971).

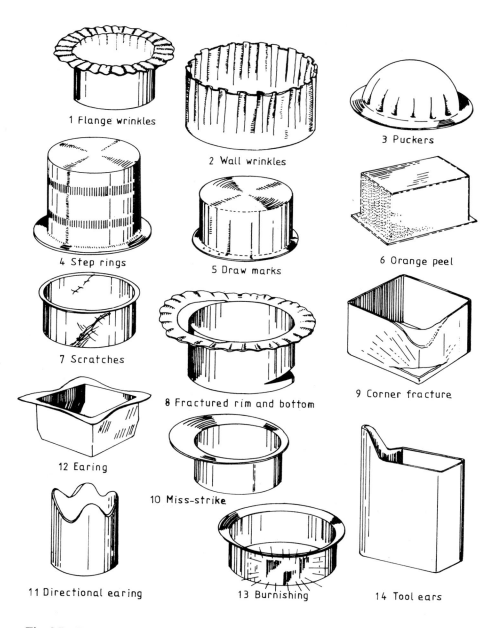

Fig. 8.7 Types of drawing defect. (After D. F. Eary and E. A. Reed, *Techniques of Pressworking Sheet Metals*. Prentice-Hall, London, 1974; also after Kaiser Aluminium.)

1 Flange wrinkles

2 Wall wrinkles

3 Puckers

4 Step rings

5 Draw marks

6 Orange peel

7 Scratches

8 Fractured rim and bottom

9 Corner fracture

12 Earing

10 Miss-strike

11 Directional earing

13 Burnishing

14 Tool ears

The geometry of the plastically deforming zone as well as the level of hydrostatic stress play important roles in workability. Backofen (1972) used a parameter Δ, given by the ratio of the material thickness to the punch or tool contact length. For the special case of frictionless plane-strain compression the dimensionless pressure $P/2k$ is shown as a function of the geometrical parameter Δ in Fig. 8.8, which also shows typical values of Δ for a range of forming processes. As the parameter Δ changes, the hydrostatic stress and the inhomogeneity of plastic deformation change. As well as larger normal interfacial pressures, for large values of Δ the deformation tends to be inhomogeneous, with hydrostatic tensions along the centreline. Inhomogeneous deformation can lead to bands of intense shear deformation, for example those associated with the dead-metal caps that appear in extrusion. In turn, zones of intense plastic deformation can be areas of premature failure.

It was observed by Rowe (1977) that it is difficult to measure workability except in terms of the actual process and work material. So the use of any model test, for example tensile or torsion tests, as a measure of workability is potentially dangerous. Despite this, empirical relationships between the model test and the process are used in practice.

Latham (1963) and Latham and Cockcroft (1966) suggested that the workability of a material is a function of the material itself as well as a function of the process. Thus

$$W = f_{1(\text{material})} . f_{2(\text{process})},\tag{8.1}$$

where f_1 is a fundamental measure of the ductility of the work material and f_2 must be a function of the process, friction and the material.

This relation for the workability was arrived at after the completion of a large number of experiments. For example, it was found that there were linear relationships between the reduction in thickness in rolling for the onset of edge cracking and the reduction in area in tensile tests. As shown in Fig. 8.9, the linear relation for initially square-edged strip has a greater slope than that for round-edged strip. Latham argued that this is caused by early inducement of tensile stress along the round edges leading to earlier fracture. Also, Latham found that there was a linear relation between the fracture strain in compression (by barrel cracking) and the tensile reduction in area, shown in Fig. 8.10.

One of the major questions in all work of this type is what is a valid fundamental measure of true or inherent ductility that can be used to define f_1? Latham noted that, because of the difficulty of necking, tension was really not suitable. Because there are no geometrical changes in a torsion test, torsional ductility, that is the shear strain at fracture, may be a more meaningful measure of f_1 than tension tests.

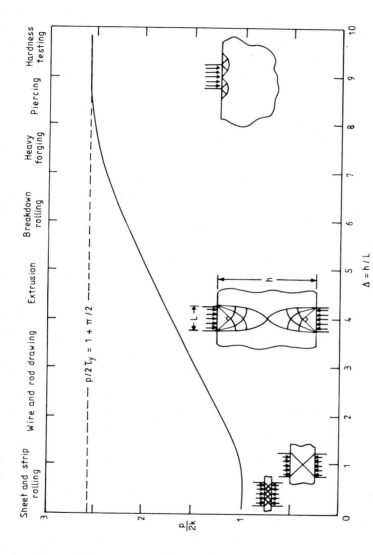

Fig. 8.8 Dependence of dimensionless yield pressure of the deformation-zone geometry for frictionless plane-strain indentation of a rigid ideally plastic solid. (After W. A. Backofen, *Deformation Processing.* Addison-Wesley, Reading, Massachusetts, 1972.)

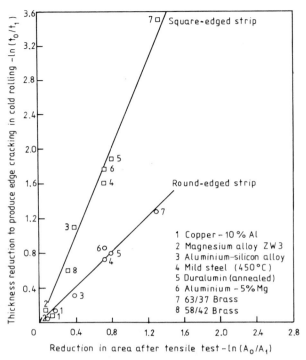

Fig. 8.9 Thickness reduction for the onset of edge cracking in cold rolling versus reduction in area in tensile tests. (After D. J. Latham and M. G. Cockcroft, NEL Report No. 216 (1966).)

Although Latham's criterion has been shown to be fairly good for predicting workability in some cold-forming processes, it has no universal applicability. Further, for hot-working processes there are complications with hot shortness.

As we have seen, friction is of fundamental importance in workability, as its presence or absence can change the mode of plastic deformation dramatically. For example, in a well-lubricated process the inhomogeneous deformation is minimized, whereas under rough conditions loads and die wear increase and bands of intense shear either appear or increase in severity if already present. One of the best illustrations of the effect of friction is on the grid distortions in forward extrusions, as shown in Fig. 8.11.

Figure 8.11(a) shows deformation confined to the dies. Figure 8.11(b, c) show the effect of increasing wall friction on the flow pattern. In Fig. 8.11(c) long wide shear bands or zones are present, but in Fig. 8.11(b) the shear bands are confined to the areas adjacent to the dead-metal cap.

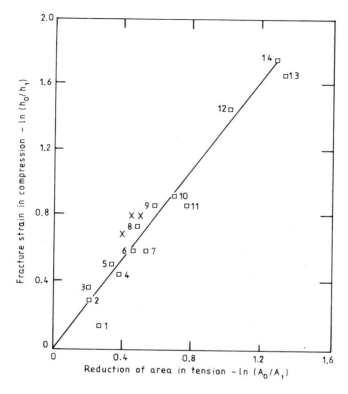

NOTATION

1 Magnesium alloy Z W 3	8 Aluminium – silicon alloy
2 Aluminium – silicon alloy (cast)	(extruded)
3 β Brass (meta)	9 Aluminium – 3% Mg
4 50/50 Brass	10 Aluminium – 5% Mg
5 58/42 Brass	11 Aluminium alloy DTD 5064
6 60/40 Brass	12 Duralumin S (annealed)
7 Duralumin S (quenched)	13 En 2A, Monel, Copper
	14 63/37 Brass

X Aluminium alloy HD33 (prestrained)

Fig. 8.10 Linear relation between fracture strain in cold compression versus reduction in area in tensile tests. (After D. J. Latham and M. G. Cockcroft, NEL Report No. 216 (1966).)

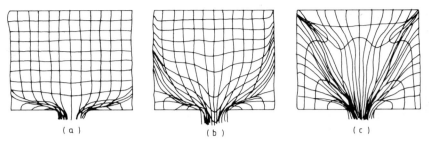

Fig. 8.11   Types of grid deformation in forward extrusion: (a) little or no wall friction; (b) moderate wall friction, showing development of dead metal cap and wide shear bands; (c) with very high wall friction deformation becomes highly distorted with complex shearing patterns. (After Pearson, *The Extrusion of Metals*, Chapman and Hall, London, 1953).

It must be concluded that there can be no universal measurement that reflects the workability of a material. But importantly, for certain processes such as cold rolling, it has been found that there is an empirical linear relation between the reduction in thickness for the onset of edge cracking and the tensile reduction in area. It is interesting to note that, although the work of Latham (1963) is quite old, it contains a very large quantity of useful experimental data and is still the starting point for researchers interested in studying workability.

8.4   Observations of workability limits

In the last section we have described some of the workability limits such as internal damage, surface defects and fittability that limit the plastic deformation. In engineering practice, all of these, especially visible defects, provide an easy and helpful way to establish the workability of materials in certain processes. Table 8.1 lists a large variety of defects observed in numerous processes; even so, the table is by no means exhaustive. Nevertheless, it presents a general picture of the range of possible defects that can occur in processed materials and are therefore related to workability.

Although defects in themselves set a visible limit on workability, their appearance is not an accurate guide to workability criteria. A limit set on plastic deformation is preferable, such as limiting plastic work per unit volume of the workpiece. Difficulties arise in attempting to establish such deformation limits by means of a continuum analysis of ductile fracture, since void nucleation and growth must depend upon the stress and strain history, i.e. the process is usually path-dependent. To overcome this difficulty, empirical criteria have been suggested for many different processes, and they have been used, often with significant success in industry.

8.4.1 Free-surface ductility in upsetting

Upsetting is the axial compression of a workpiece in order to enlarge the cross-sectional area of the whole or part of its length. Two possibilities are shown in Figs. 8.12(a, b). A third possibility, called heading, is one in which one end of the workpiece is enlarged as shown in Fig. 8.12(c).

The workability of a material in upsetting, sometimes known as the free-surface ductility or upsettability of the material, is dependent on the states of stress and strain on the expanding free surface. For example, if tensile stresses are induced at the free surface by high frictional constraint at the die–workpiece interfaces; this can lead to extensive barrelling and early fracture at the equator. On the other hand, well-lubricated compression will produce insignificant tensile stresses, and the surfaces will expand uniformly, suppressing cracking completely, at least theoretically.

A number of other variables have been found to affect upsettability; these are specimen aspect ratio (height–diameter ratio), amount of prestrain in the specimen and the nature of the free surface—there may be grooves or gouges left from prior processing such as wire drawing.

In the review by Jenner and Dodd (1981) of upsetting and free-surface ductility it was noted that Latham (1963) was apparently the first to report two modes of fracture in cylindrical specimens. For ductile materials fracture was confined to the barrel, whereas for less-ductile materials fracture occurs catastrophically through the entire body of the workpiece. This latter behaviour has been confirmed more recently by Dodd et al. (1985) for a magnesium alloy and Clift (1986) for 60–40 brass. By using the apparent disadvantage of barrelling, Kudo and Aoi (1967) and Kudo et al. (1968), developed a linear fracture condition in strain space for a medium-carbon steel (S45C). Kudo et al. measured strain histories to fracture at the equatorial free surfaces of specimens using identation marks. Cylindrical specimens with aspect ratios varying between $\frac{1}{2}$ and 2 were used. Figure 8.13 shows the results for flat and concentrically grooved dies. Lubricated flat dies lead to long strain paths to fracture, whereas concentrically grooved dies lead to steep strain paths and early fracture. Kudo et al. suggested that the fracture condition in their upsetting tests is given by the straight-line condition

$$\varepsilon_{\theta f} = a - \tfrac{1}{2}\varepsilon_{zf} \tag{8.2}$$

where $\varepsilon_{\theta f}$ is the hoop fracture strain, ε_{zf} is the axial fracture strain and a is the

Table 8.1

The common principal defects associated with various metal-forming processes

ROLLING	FORGING
Flat and section rolling	*Open- and closed-die forging, upsetting, indentation*
Edge cracking	Longitudinal cracking
Transverse—fire cracking	Hot tears and tears
Alligatoring (crocodiling)	Edge cracking
Fishtail	Central cavity
Folds, laps	Centre bursts
Flash, fins	Thermal cracks
Laminations	Folds, laps
Ridges—spouty material	Flash, fins
Ribbing	Laminations
Sinusoidal fracture	Orange peel
	Shearing fracture
Cross transverse and helical rolling	Piping
Central cavity (axial or annular fissure)	
Overheated ball bearing	*Deep-drawing*
Roll mark	Wrinkling
Folding (or seaming) laps, fringes	Puckering
Squaring	Tearing (necking)
Necking	Edge cracking
Triangulation and triangular fishtail	Orange peel
	Stretcher strains (Lüders lines)
Ring rolling	Earing
Cavities	*Bending and contour forming*
Fishtail	Cracking
Edge cracking	Wrinkling
Straight-sided forms	Springback

plane-strain fracture strain. The fracture line is parallel to the line for homogeneous compression, for which Fig. 8.14 illustrates the two types of fracture observed by Kudo and Aoi. They found that the fracture coincides with the maximum-shear-stress plane.

Subsequently other researchers including Kobayashi (1970), Lee (1972) and Lee and Kuhn (1973) have confirmed the straight-line fracture condition found by Kudo and Aoi.

The straight-line fracture condition for surface cracking in upsetting and related processes is an important guide to the workability a material. Once the diagram has been determined experimentally for a particular material, then clearly, for crack-free upsetting, surface strain paths should always be

Hole Flanging
Lip formation
Petal formation
Plug formation

Rotary Forging
Mushrooming:
  Central fracture
  Flaking

High-energy-rate forging
Piping
Dead-metal region
Laps
Turbulent metal flow

Extrusion Piercing
Christmas tree (fir tree):
  Hot cold shortness
  Radial and circumferential cracking
  Internal cracking
  Central burst (chevrons)
  Piping (cavity formation)
  Sucking in
Corner lifting
  Skin inclusions (side and bottom of the billet)
  Longitudinal streaks
  Laps
Laminated fractures:
  Mottled appearance
  Extrusion defect

Impact extrusion
Multiple tensile "necks"
Thermal breakoff

Drawing of Rod, Sheet Wire and Tube
Internal bursts (cup and cone, chevron)
Transverse surface cracking
Chips of metal
Poor surface finish
Folding and buckling
Fins, laps
Chatter (vibration) marks
Season cracking
Island-like welding

Blanking and cropping
Distortion of the part
(doming and dishing)
Cracking
Martensitic lines
Eyes, ears, warts, beards, tongues

Spinning, flow turning, shear forming
Springback
Wall-fracture (shear splitting and circumferential splitting)
Wrinkling
Buckling
Back-extrusion (overreduction)
Under-reduction

Peen forming, ball forming
Overlapping dimples
Orange peel
Surface tearing
Break-up of surface grains
Intergranular cracking
Wrinkling
Microfissures
Folds

From W. Johnson and A. G. Mamalis, in *Proc. 17th Int. Machine Tool Design Research Conf.* (ed. S. A. Tobias), pp. 607–621. Macmillan, London, 1977.

below the fracture line. To avoid fracture when the strain path of a material crosses the fracture line, either the material should be changed or the strain path made less steep by better lubrication.

de Meester and Tozawa (1979) have now proposed a standard test for upsettability. Two cylindrical specimens are suggested, as shown in Fig. 8.15.

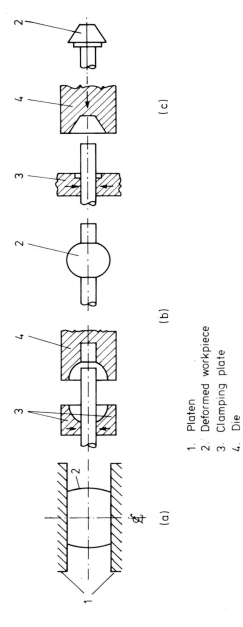

Fig. 8.12 (a) Upsetting of the whole of the length of a workpiece; (b) upsetting part of the length of a workpiece; (c) heading.

1. Platen
2. Deformed workpiece
3. Clamping plate
4. Die

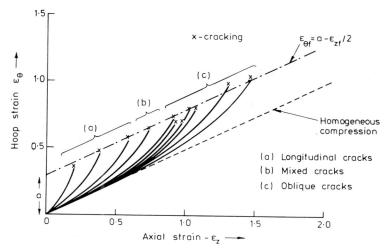

Fig. 8.13 Straight-line fracture condition for a medium-carbon steel observed by H. Kudo and K. Aoi (*J. Japan Soc. Tech. Plasticity* **8**, 17–27 (1967)).

The plane cylindrical specimen is for assessment of the conventional upsettability, while the grooved specimen is for the assessment of the effects of longitudinal grooves on the upsettability. Each specimen has central cones at each end, which locate the specimen on the concentrically grooved dies, as shown in Fig. 8.16. When cracks of lengths greater than half a millimetre are observed in the test, this point is regarded as the upsettability limit.

In detailed fine-grid upsetting work by Lee (1972), a strain perturbation was observed just prior to rapid void growth. This perturbation, equivalent do $d\varepsilon_z = 0$, as shown in Fig. 8.17, corresponds to a transient plane-strain stress state at the surface just prior to fracture, and is certainly a more conservative upsettability limit.

8.4.2 Limitations in sheet-metal forming

For sheet forming, the major limitation is localized necking, as we have already described. The onset of localized necking corresponds to a bifurcation in plastic flow. Eventually fracture occurs in the more extensively strain-hardened material that constitutes the neck.

Nowadays, it is common practice to represent the onset of localized necking under different strain ratios in terms of the major and minor in-plane strains ε_1 and ε_2 (or in terms of engineering strains). These loci are called forming-limit diagrams, and two examples for aluminium are shown in Fig. 8.18. It can be seen that out-of-plane stretching normally produces a higher

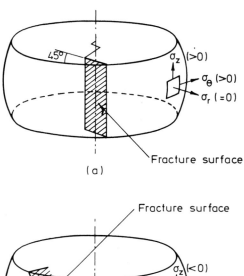

(a)

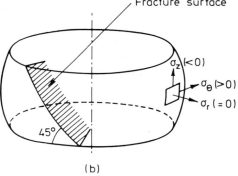

(b)

Fig. 8.14 Longitudinal and diagonal fracture planes observed by H. Kudo and K. Aoi (*J. Japan Soc. Tech. Plasticity* **8**, 17–27 (1967)).

forming-limit diagram than in-plane stretching. Hecker (1978) has used both out-of-plane and in-plane stretching tests extensively. The punch stretching apparatus is shown in Fig. 8.19(a). An important feature of punch stretch tests is that strain gradients are always present across the dome shape. On the other hand, using Hecker's in-plane stretching apparatus (which is based on a test described by Marciniak and Kuczynski (1967)), shown in Fig. 8.19(b), no such strain gradients are present. Here a washer with a circular hole in its centre is used between the punch and the specimen, and the punch has a circular cutout in its flat nose. In both punch stretching and in-plane stretching the strain ratios were varied by varying the widths of the specimens; the amount of drawing in of the specimen being controlled by its width.

Although the workability limits in sheet forming are determined by the

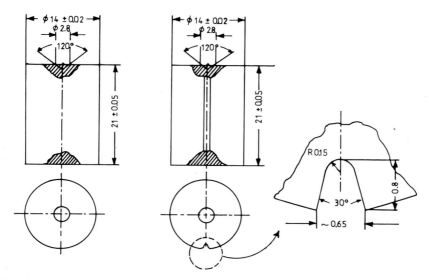

Fig. 8.15 Two standardized specimens for upsetting tests: (a) conventional cylindrical specimen; (b) longitudinally grooves specimen. (After B. de Meester and Y. Tozawa, *Ann. CIRP* **28**, 577–580 (1979).)

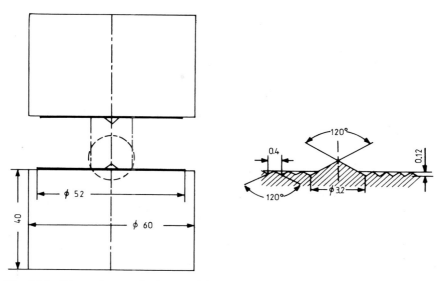

Fig. 8.16 Die set for standard tests. (After B. de Meester and Y. Tozawa, *Ann. CIRP* **28**, 577–580 (1979).)

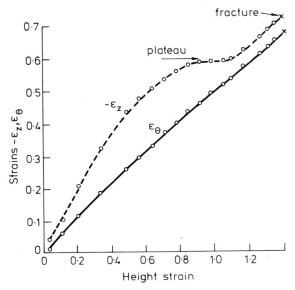

Fig. 8.17 Strain perturbation, corresponding to $d\varepsilon_z = 0$, observed by Lee (1972). After this perturbation there is a sudden void coalescence at the cylinder equator.

onset of localized necking, if stretching is continued the fracture locus is often found to be approximately linear with a slope of $\tan(-1)$, as described by Atkins (1980a) (see Fig. 8.20) for aluminium alloys. This corresponds to an empirical fracture criterion which is equivalent to a constant-thickness criterion for the onset of fracture.

Any complete theory of ductile fracture must be able to explain and differentiate between the upsetting line with a slope of $\tan^{-1}(-\frac{1}{2})$ and the sheet-forming locus.

8.5   Empirical criteria for workability

Numerous attempts have been made to relate the fracture strains of metals to macroscopic variables associated with the metal, the process, or both. Such criteria, although perhaps only applicable to one process or class of processes, have obvious advantages in industry. Even if a different criterion is required to determine the workability of a metal in different processes, these relatively simple criteria may be easier to manipulate than a universal ductile-fracture criterion. An initial choice for a macroscopic criterion is one based

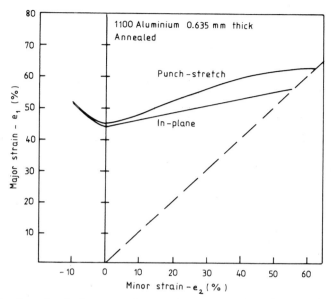

Fig. 8.18 Forming-limit curves for 1100 aluminium measured both by punch stretching and in plane stretching. (After S. S. Hecker, in *Applications of Numerical Methods to Forming Processes* (ed. H. Armen and R. F. Jones), pp. 85–94. ASME AMD-28, 1978.)

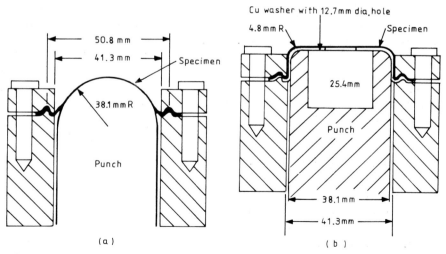

Fig. 8.19 (a) Punch stretching apparatus and (b) punch apparatus used for in-plane tests by S. S. Hecker (in *Applications of Numerical Methods to Forming Processes* (ed. H. Armen and R. F. Jones), pp. 85–94. ASME AMD-28, 1978).

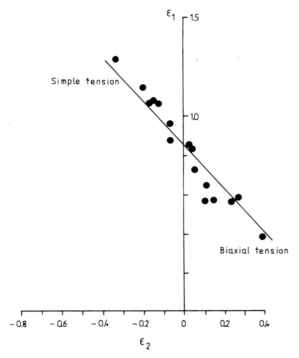

Fig. 8.20 Fracture points for sheet testing. (Based on a diagram taken from A. G. Atkins, *The Mechanics of Ductile Fracture in Metalforming*. ICFG, 1980.)

on the total plastic work done up to fracture. This could be stated as "fracture will occur when the total plastic work per unit volume reaches some characteristic critical value". Assuming that the generalized von Mises stress and strain apply to the metal, the plastic work criterion can be written as

$$\int_{0}^{\bar{\varepsilon}_f} \bar{\sigma}\, d\bar{\varepsilon} = A, \tag{8.3}$$

where $\bar{\varepsilon}_f$ is the effective fracture strain and A is a constant for the metal and a particular process.

Latham (1963) observed that fractures in metalworking processes tend to occur in the region of largest tensile stress, and he proposed the following criterion of critical tensile plastic work:

$$\int_{0}^{\bar{\varepsilon}_f} \sigma_T\, d\bar{\varepsilon} = B, \tag{8.4}$$

where σ_T is the largest tensile stress and B is a constant characteristic of both the material and the process.

A modification of the Latham criterion was suggested by Brozzo *et al.* (1972); this modification includes a hydrostatic-pressure term

$$\int_0^{\bar{\varepsilon}_f} \frac{2\sigma_T \, d\bar{\varepsilon}}{3(\sigma_T - \sigma_m)} = C, \qquad (8.5)$$

where σ_m is the hydrostatic stress and C is dependent on the metal and process as before.

Clift *et al.* (1985) have carried out some plane-strain side-pressing experimentation on an aluminium alloy. Also, in parallel with this a finite-element model of the process was run. The finite-element results were used with the above criteria to predict the site of fractures observed experimentally. It was found that the total plastic-work criterion was the only successful

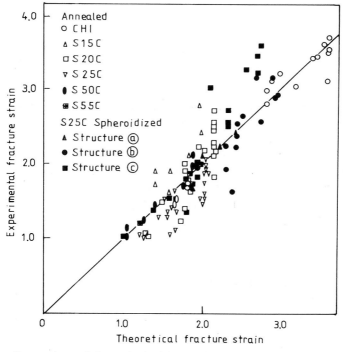

Fig. 8.21   Comparison of theoretical with experimental fracture strains for steels in torsion under various hydrostatic pressures. (After K. Osakada *et al.*, *Bull. JSME* **21**, 1236–1243 (1978).)

criterion for the prediction of the site of fracture and the amount of plastic deformation.

A valuable semi-empirical criterion has been developed by Osakada *et al.* (1977, 1978, 1981). The criterion has the form

$$\int_0^{\bar{\varepsilon}_f} \langle \bar{\varepsilon} + ap - b \rangle \, \mathrm{d}\bar{\varepsilon} = c, \tag{8.6}$$

where a, b and c are constants that are derived from experiments, and the function $\langle x \rangle$ is defined by

$$\langle x \rangle = \begin{cases} x & (x > 0), \\ 0 & (x < 0). \end{cases}$$

This criterion was derived after a large number of tension, compression and torsion tests on carbon steels. The criterion can also take account of superimposed hydrostatic pressure through the term p and prestrain. The constants a and c appear to be the same for different steels, but it is found that the value of b is affected by microstructural factors such as size, shape and volume fracture of cementite. Figure 8.21 shows the experimental fracture strains for various steels compared with the theoretically derived ones. The experimental points are for torsion tests with different superimposed hydrostatic-pressure histories.

Comparison of these empirical criteria with the theoretical ones in §5.2.2 concerning void growth shows that both groups of criteria emphasize the importance of hydrostatic stress on ductile fracture. However, there is a great gap between their quantitative forms.

9

Workability in Sheet-Metal Forming

9.1 Characteristics of sheet-metal forming

Sheet-metal forming processes such as stretch forming, pressing, deep drawing, ironing and bending are not usually limited by fracture, but rather by strain localization along one or more fairly narrow bands. These bands grow into sharp necks, along which ductile fracture can eventually occur. Therefore for sheet forming, workability is normally governed by the stability of deformation. To improve workability in these circumstances, then, it is necessary to delay until larger strains a bifurcation of plastic flow to a localized mode.

The various sheet-metal shapes were classified by Sachs (1951) into the following five broad categories (see Fig. 9.1):

(1) singly curved parts, for example straight sections, straight-flanged parts and single-curvature contoured parts;
(2) contoured flanged parts, including stretch and shrink flanges;
(3) curved sections, including sections with uniform and nonuniform curvature, compact curved sections;
(4) deep recessed parts with vertical and sloping walls, including cups, tubes and box shapes and semitubular parts;
(5) shallow recessed parts, including parts with double curvature, dish shapes, beaded, embossed and corrugated parts.

As well as localized necking, another form of failure is wrinkling or buckling, in, for example, the flanges of deep-drawn cups due to the use of too low blank-holding force. The flanges of cups are subjected to compressive forces, and as the onset of buckling is dependent upon the elastic modulus

207

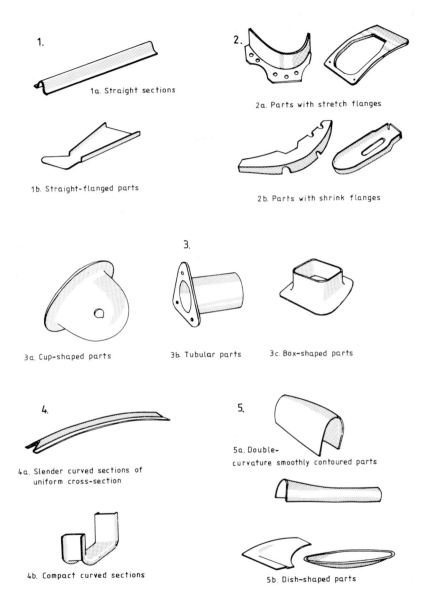

1.

1a. Straight sections

1b. Straight-flanged parts

2.

2a. Parts with stretch flanges

2b. Parts with shrink flanges

3.

3a. Cup-shaped parts

3b. Tubular parts

3c. Box-shaped parts

4.

4a. Slender curved sections of uniform cross-section

4b. Compact curved sections

5.

5a. Double-curvature smoothly contoured parts

5b. Dish-shaped parts

Fig. 9.1 Classification of sheet-metal components by G. Sachs (*Principles and Methods of Sheet-Metal Fabricating.* Reinhold, New York, 1951).

and the flange thickness, it can only be overcome by increasing the blank-holding force. The possible modes of failure can be predicted from the stress state imposed on the sheet. According to Johnson (1972b), wrinkling can occur in regions of high compressive stress, zone A in Fig. 9.2. Local tensile instability occurs in regions where a line of zero extension can exist, regions B and D, and biaxial tensile failure occurs in region C (see below).

The bending of sheets and other sections is one of the most common processes. In sheet bending, a flat sheet is formed into a curved shape by one of a number of different possible methods, for example simple hand bending or die pressing. It is found in bending operations that, for a particular section, there is a minimum bend radius beyond which necking and fracture can occur. This characteristic radius is clearly the forming limit in bending processes. The minimum bend radius is usually presented in terms of the ratio of the minimum bend radius to the metal thickness. Important factors that influence this ratio are the workpiece width and edge finish.

A second problem that can arise in bending operations is elastic springback. In this case, under load the inside surface of the workpiece matches the contour of the die, but when the load is released the workpiece radius becomes greater. Allowance for elastic springback must be made in die design, and this aspect of formability becomes very important in complex components. Thus the formability of large pressings, for example car body panels, is dependent not only on the occurrence of strain localization, but also

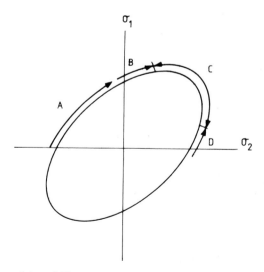

Fig. 9.2 Modes of instability, B and D tensile instability and C biaxial tensile instability. (After W. Johnson, *Metallurgia and Metal Forming* **39**, 89–99 (1972).)

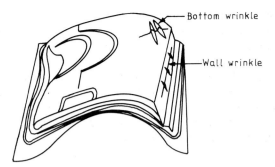

Fig. 9.3 Typical wrinkles in an automobile front fender. (After T. Furabayashi and M. Kojima, in *Formability of Metallic Materials—2000 A.D.* (ed. J. R. Newby and B. A. Niemeier). ASTM STP 753 (1982).)

on the degree of surface distortions of the panel, such as warping, wrinkling and elastic deflections. The elastic deflections are caused by springback after unloading. The mechanical properties of sheets affecting the fitting behaviour correspond to the fittability of the material (Ishigaki, 1978). A minor improvement in the fitting behaviour of a sheet can decrease the scrap rate markedly because of the high volume production of these pressings, and can therefore lead to significant financial saving.

A diagram of typical surface defects is shown in Fig. 9.3. With gridding, deflections and the origin of defects can be seen more easily. The die face is one of the most important variables in determining formability, product quality and scrap rate.

9.2 Plastic anisotropy

9.2.1 Anisotropic yield criteria

In sheet metals, plastic anisotropy plays an important role in determining forming limits. During plastic deformation of metals a preferred orientation of the grains develops. The preferred orientation is the reason for the development of plastic anisotropy in metals. Plastic anisotropy is often shown to a marked degree in rolled sheet, in which the yield stress in the rolling direction is not equal to that in the transverse direction, i.e. the plastic properties vary with orientation in the plane of the sheet (planar anisotropy). Also, the mechanical properties in the plane of the sheet generally differ from the through-thickness properties (normal anisotropy).

Hill (1948) proposed an anisotropic yield criterion to take account of the variations in mechanical properties with direction. In this quadratic criterion it is assumed that the anisotropy has three mutually orthogonal planes of symmetry; the intersections of these planes are referred to as the principal axes of anisotropy. We assume that the x-, y- and z-directions coincide with the rolling, transverse and through-thickness directions of the sheet respectively. These assumptions are weak with respect to anisotropy of crystallographic origin. It is further assumed that the Bauschinger effect is absent. Then Hill's yield criterion can be written as

$$2f(\sigma_{ij}) = F(\sigma_y - \sigma_z)^2 + G(\sigma_z - \sigma_x)^2$$
$$+ H(\sigma_x - \sigma_y)^2 + 2L\tau_{yz}^2 + 2M\tau_{zx}^2 + 2N\tau_{xy}^2$$
$$= 1, \tag{9.1}$$

where F, G, H, L, M and N are parameters that are characteristic of the current anisotropy. It should be noted that this criterion reduces to the von Mises criterion when the material is isotropic. If we assume that (9.1) is the plastic potential then the plastic strain increments can be shown, by partial differentiation, to be given by

$$\left.\begin{aligned}
d\varepsilon_x &= d\lambda\,[H(\sigma_x - \sigma_y) + G(\sigma_x - \sigma_z)], \\
d\varepsilon_y &= d\lambda\,[F(\sigma_y - \sigma_z) + H(\sigma_y - \sigma_x)], \\
d\varepsilon_z &= d\lambda\,[G(\sigma_z - \sigma_x) + F(\sigma_z - \sigma_y)], \\
d\gamma_{yz} &= d\lambda\,L\tau_{yz}, \\
d\gamma_{zx} &= d\lambda\,M\tau_{zx}, \\
d\gamma_{xy} &= d\lambda\,N\tau_{xy},
\end{aligned}\right\} \tag{9.2}$$

since the principal axes of stress coincide with those of strain increment and anisotropy. Volume constancy is preserved during plastic deformation; thus

$$d\varepsilon_x + d\varepsilon_y + d\varepsilon_z = 0. \tag{9.3}$$

If we confine attention to the plane-stress case, with the out-of-plane stress σ_z being zero, then

$$(G + H)\sigma_x^2 - 2H\sigma_x\sigma_y + (H + F)\sigma_y^2 + 2N\tau_{xy}^2 = 1. \tag{9.4}$$

9.2.2 The strain ratio r

The r-value is a particularly valuable measure of anisotropy in sheet metals. The strain ratio is defined as the ratio of the width to the thickness strain in a

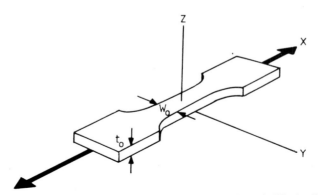

Fig. 9.4 Strip tensile specimen. (After W. F. Hosford and W. A. Backofen, in *Fundamentals of Deformation Processing: Proc. 9th Sagamore Army Materials Research Conf.* (ed. W. A. Backofen *et al.*). Syracuse University Press, New York, 1964).

uniaxial tensile test. Figure 9.4 shows a uniaxial tensile specimen from which the strain ratio can be measured. The r-value is defined as

$$r = \varepsilon_w / \varepsilon_t \qquad (9.5)$$

where $\varepsilon_w = \ln(w/w_0)$ and $\varepsilon_t = \ln(t/t_0)$.

For an isotropic material the r-value is unity. A *large r-value implies a resistance to thinning*, and conversely a low r-value implies a low through-thickness strength.

It is usual for the r-value to vary with direction in the plane of the sheet; however, for mathematical simplicity the assumption is made that the r-value is constant within the plane, i.e. the material is assumed to be isotropic in the plane. It is common to characterize such a material by $\bar{r}$, where

$$\bar{r} = \tfrac{1}{4}(r_0 + 2r_{45} + r_{90}), \qquad (9.6)$$

the subscripts referring to the tensile loading angles from the rolling direction. Also, it can be shown easily that

$$r_0 = \frac{H}{G}, \quad r_{90} = \frac{H}{F},$$

so if the material is assumed to be isotropic in the plane then $r_0 = r_{90}$ and $G = F$.

9.2.3 Anomalous behaviour

Recently interest has been revived in yield criteria and sheet-metal formability in relation to so-called anomalous materials. Woodthorpe and Pearce

(1970) found that the r-value was about 0.5 for commercial-purity aluminimum, yet the ratio of the yield strength in balanced biaxial tension, σ_b, to that in uniaxial tension, σ_u, was greater than unity. Using Hill's (1948) criterion for a plant isotropic material and taking ratios,

$$\left(\frac{\sigma_b}{\sigma_u}\right)^2 = \frac{1+r}{2}. \tag{9.7}$$

It is clear that $(\sigma_b/\sigma_u) > 1$ if $r > 1$ and $(\sigma_b/\sigma_u) < 1$ if $r < 1$. However, since for the material tested by Woodthorpe and Pearce $\sigma_b/\sigma_u > 1$ when $r < 1$, these authors termed this "anomalous behaviour". To encompass this anomaly in aluminium and other materials, Hill (1979) proposed the following nonquadratic anisotropic criterion:

$$f|\sigma_2 - \sigma_3|^m + g|\sigma_3 - \sigma_1|^m + h|\sigma_1 - \sigma_2|^m + a|2\sigma_1 - \sigma_2 - \sigma_3|^m$$
$$+ b|2\sigma_2 - \sigma_3 - \sigma_1|^m + c|2\sigma_3 - \sigma_1 - \sigma_2|^m = \sigma^m, \tag{9.8}$$

where the six coefficients f, g, h, a, b and c characterize the anisotropy; σ is a scaling factor for stress, and $m > 1$ to insure convexity of the yield locus. If $m = 2$ and the coefficients are rearranged then this criterion reduces to Hill's quadratic criterion. Four special cases of (9.8) have been proposed by Hill:

(i) $a = b = h = 0$, $f = g$;
(ii) $a = b$, $c = f = g = 0$;
(iii) $a = b$, $f = g$, $c = h = 0$;
(iv) $a = b = f = g = 0$.

The general expressions for r and the ratio σ_b/σ_u are

$$r = \frac{a(2^{m-1} + 2) - c + h}{a(2^{m-1} - 1) + 2c + f}, \tag{9.9}$$

$$\left(\frac{\sigma_b}{\sigma_u}\right)^m = \frac{1+r}{2}\left\{1 + \frac{(2^{m-1} - 2)(a - c)}{a + c2^{m-1} + f}\right\}, \tag{9.10}$$

assuming planar isotropy.

Dodd and Caddell (1983) have investigated the ranges of application of each of these four special cases of this non-quadratic yield criterion. As Parmar and Mellor (1978) discovered, case (iv) of the criterion is simple to utilize. More recently, Budiansky (1984) introduced a general yield criterion applicable for plane-stress conditions and materials having planar isotropy. Budiansky showed that a special case of his criterion is the same as Hill's (1979) case (iv) criterion.

9.3 Forming limits

In many sheet-metal forming operations the deformation is predominantly a stretching. When a sheet is progressively thinned two modes of plastic instability are possible. These are diffuse and localized necking (see §§2.1 and 7.1). We have already seen that diffuse necking occurs at a peak in the load in a tensile test on a sheet specimen. This was generalized by Swift (1952) to allow for various stress ratios. Diffuse necking occurs in annealed sheets. In cold-worked sheets localized necking occurs and diffuse necking can be undetected. There are basically two different modes of localized neck formation in sheet metals. If one of the in-plane strains is compressive then the angle of the localized neck will depend on the imposed stress ratio. In this case the angle of the neck with respect to the pulling axis is a characteristic angle. On the other hand, if both in-plane strains are tensile then a localized neck forms whose length is usually perpendicular to the direction of the larger stress.

Forming-limit diagrams show the combination of major and minor in-plane strains beyond which failure occurs; failure usually corresponds to localized neck formation. Keeler and Backofen (1963) seem to have been among the first researchers to produce forming-limit diagrams for localized thinning. They found that in biaxial tension the larger principal strain increases with increasing biaxiality.

Further work by Keeler (1965, 1968) and Goodwin (1968) developed forming-limit diagrams for carbon-steel stampings, and as a consequence these curves are often referred as Keeler–Goodwin diagrams.

A typical forming-limit diagram is shown in Fig. 9.5(a), which is for a carbon steel. Figure 9.5(b) shows, for aluminium-killed steel, that, irrespective of strain ratio applied by the test, the strain path tends to plane strain as it approaches the forming-limit curve, i.e. the strain paths become vertical with $d\varepsilon_2 = 0$. The level of the forming-limit diagram is sensitive to the strain hardening exponent n. The greater the n-value, the higher the limit diagram.

As was observed by Korhonen (1981), necking is associated with the end of uniqueness and stability of plastic deformation. The end of uniqueness corresponds to a point where a number of solutions can simultaneously satisfy the boundary conditions. Hill (1957) stated that bifurcation of flow into the states of equilibrium infinitesimally close to the original can occur before a failure in stability. Bifurcation is associated with the end of uniqueness of the deformation and may precede stability loss (details are given in §§7.2.4 and 7.2.5).

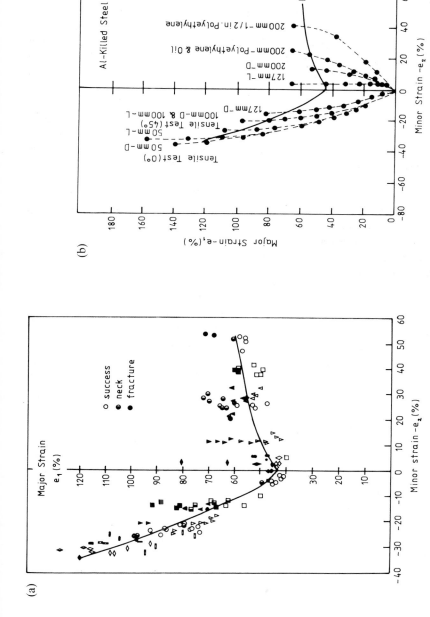

Fig. 9.5 (a) Forming-limit diagram for low-carbon steel (plotted in terms of engineering strains). (b) The relation of the strain path to the FLC of Al-killed steel. (After S. S. Hecker, in *Formability: Analysis, Modelling and Experimentation* (ed. S. S. Hecker *et al.*), pp. 150–182. Met. Soc. AIME, 1977.)

9.3.1 Effect of loading method on forming limits

Essentially, there are two different methods of testing a sheet of metal to its forming limits; these are in-plane stretching and out-of-plane stretching with a punch or hydraulic pressure. For in-plane stretching the stresses are applied by a die with an orifice; the sheet is stretched across the orifice and is subjected to a fairly uniform stress state. One method for out-of-plane testing is to use a hemispherical punch, which is gradually forced against the sheet. In this case the stress and strain states vary according to position on the surface.

In hydraulic bulging the greatest thinning occurs at the pole of the stretched cup. However, as observed by Hecker (1977) in punch stretching, strain concentration and fracture have been observed between the pole and clamp. The meridional strain goes through a peak between these two positions.

It was observed by Azrin and Backofen (1970), Hecker (1978) and others that the strain gradient in the region of the neck is significantly greater for in-plane than out-of-plane stretching. It is generally observed that punch stretching and hydraulic bulging of sheets produce a more gradual change in strain path in the zone of fracture. Figure 9.6 shows the strain paths of the fracture area for in-plane and bulge tests on aluminium-killed steel. However,

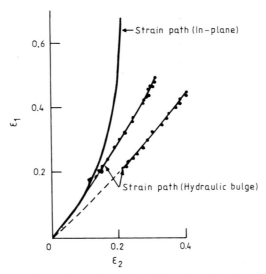

Fig. 9.6 Differences in strain paths at the failure location for in-plane stretching (Azrin and Backofen, 1970) and bulging (Painter and Pearce, 1974) for killed steel. (Adapted from A. K. Ghosh, in *Mechanics of Sheet Metal Forming* (ed. D. P. Koistinen and N.-M. Wang), pp. 287–312. (Plenum Press, New York, 1978.)

it should be noted that for some materials in-plane and hydraulic tests give very similar limit strains.

It is clear that the differences between the forming limits for punch and in-plane stretching are derived from the differences in strain gradients between the two tests caused by differences in test geometry and the absence of any frictional contribution for in-plane stretching. Of the three test arrangements, hydraulic bulging of a diaphragm is by far the least rigid test set-up.

We have already seen in §8.4.2. that the forming-limit diagram for in-plane stretching is less extensive than that for punch stretching. This type of behaviour is clearly applicable to other metals besides the aluminium of Fig. 8.18.

9.3.2 Classification of forming limits and relevant material parameters

It is difficult to classify forming behaviour of different metals; however, Azrin and Backofen (1970) and Mellor (1982) observed distinct categories of forming-limit diagram. According to Mellor, there are three general classes of forming-limit behaviour in the biaxial tension quadrant. These are shown in Figs. 9.7(a–c); Fig. 9.7(a) for rimming steel shows a rising characteristic for $\varepsilon_2 = 0$ to $\varepsilon_1 = \varepsilon_2$. The limit-strain locus for commercial-purity aluminium shows a slight drop as the balanced biaxial point $\varepsilon_1 = \varepsilon_2$ is approached. Finally the limit strains for 70/30 brass are all below the Swift diffuse-instability locus, which is much lower than the predicted strains. Azrin and Backofen confirmed in their earlier work that steel had a rising characteristic on the forming-limit diagram. In general, high-strain hardening and high-strain-rate hardening will lead to higher forming-limit diagrams. As noted by Hecker (1977), the high-forming-limit curves of low-carbon steels result from reasonably high-strain hardening and strain-rate sensitivity whereas the lower curves for aluminium alloys are probably caused by insensitivity to strain rate. Some materials that strain-harden markedly (e.g. brass and stainless steel) still only have forming-limit diagrams no higher than those for plain carbon steels. This is because of the lack of rate sensitivity shown by the former materials. Biaxial limit strains are sensitive to microstructure, and therefore variations in microstructure should be borne in mind in relation to these generalizations of behaviour.

Hecker (1977) showed that lubrication moves the strain ratio to the right of the forming-limit diagram. Importantly, for $\varepsilon_2 > 0$ better lubrication will lead to better formability, and for $\varepsilon_2 < 0$ the reverse holds.

The effect of strain-path shape on the forming-limit diagram was discussed by Korhonen (1978). It was shown that the position of the limit curves was very sensitive to strain path. If the strain path curves towards the plane-strain conditions then this decreases the level of the whole forming-limit curve.

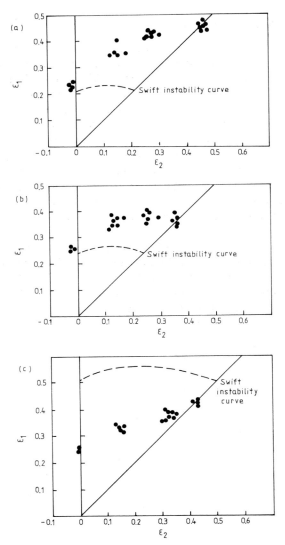

Fig. 9.7 Experimental limit strains for (a) rimming steel, (b) annealed commercial-purity aluminium and (c) 70/30 brass. (After P. B. Mellor, in *Mechanics of Materials: The Rodney Hill 60th Anniversary Volume* (ed. H. G. Hopkins and M. J. Sewell), pp. 383–415. Pergamon Press, Oxford, 1982.)

9.4 Simplified theories of localized necking

9.4.1 Local necking (Hill, 1952)

We have seen that in a uniaxial tensile test carried out on an annealed sheet specimen, diffuse necking occurs at a peak in the load (see Chapter 7). On further straining, a localized neck forms, with all the plastic strain being confined to the necked region. Finally, fracture occurs along the necked region. Consideration is confined here to the more generalized situation of sheet materials subjected to biaxial in-plane stresses. The variation in thickness is necessarily assumed to be small enough such that plane-stress conditions apply. It is shown here that a localized-necking type of discontinuity is permissible when the governing equations are hyperbolic and characteristic solutions occur. When the governing equations are hyperbolic it is shown that it is possible for a neck to form along a characteristic (see §7.4). The following basic assumptions about the material and its yield function are made: (1) the material is isotropic; (2) the yield function and the plastic potential coincide; (3) the Bauschinger effect is absent; (4) the yield function has sixfold symmetry; (5) the yield surface is everywhere convex to the origin; (6) the material is perfectly plastic. The type of necking discontinuity considered by Hill (1952) is shown in Fig. 9.8. Here v_i is the relative velocity vector of the two shoulders on either side of the neck. The

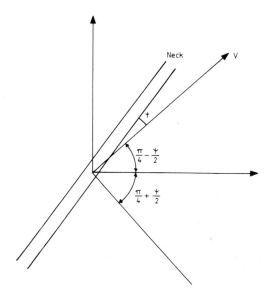

Fig. 9.8 Localized necking along a characteristic direction.

neck coincides with one characteristic and the velocity vector is normal to the other, as shown. The principal strain increments are

$$\left. \begin{array}{l} d\varepsilon_1 = \dfrac{v}{2b}(1 + \sin\psi), \quad d\varepsilon_2 = -\dfrac{v}{2b}(1 - \sin\psi), \\[2mm] d\varepsilon_3 = -\dfrac{v}{b}\sin\psi, \end{array} \right\}$$ (9.11)

where v is the magnitude of the velocity vector and b is the breadth of the neck. Also,

$$\sin\psi = \frac{d\varepsilon_1 + d\varepsilon_2}{d\varepsilon_1 - d\varepsilon_2} = \frac{1 + \dfrac{d\varepsilon_2}{d\varepsilon_1}}{1 - \dfrac{d\varepsilon_2}{d\varepsilon_1}}.$$ (9.12)

According to the definition of the principal strains, $d\varepsilon_1 > d\varepsilon_2$, and $d\varepsilon_1 > 0$ in sheet forming. Hence $d\varepsilon_2/d\varepsilon_1 > 0$ corresponds to $d\varepsilon_2 > 0$, i.e. for biaxial tension the right-hand side of (9.12) is greater than unity. In this case there is no angle ψ or characteristic solution that can satisfy (9.12). Thus local necking is limited to cases where $d\varepsilon_2 \leqslant 0$, as $d\varepsilon_2 \to 0$, $\psi \to \frac{1}{2}\pi$, namely where the local neck becomes perpendicular to the direction of major strain. Equation (9.12) can also be derived based on Mohr's circle of strain. All Mohr circles for cases $d\varepsilon_2 > 0$ are located to the right of the γ-axis (compare with Fig. 2.4); therefore there is no inextensional direction. This means that local necking is excluded when $d\varepsilon_2 > 0$.

We assume that the hardening is a function only of the plastic work, i.e.

$$f(\sigma_1, \sigma_2) = F(W_p).$$ (9.13)

It follows that

$$\left(\frac{\partial f}{\partial \sigma_1}\right) d\sigma_1 + \left(\frac{\partial f}{\partial \sigma_2}\right) d\sigma_2 = (\sigma_1\, d\varepsilon_1 + \sigma_2\, d\varepsilon_2) F'(W_p).$$ (9.14)

For localized necking it is assumed that the rate of hardening within the neck balances the rate of diminution in thickness; see (2.7) and (2.8).

Analogously to the requirement for localized necking in uniaxial tension of a sheet (§2.1.2), the condition for localized necking in a biaxial stress state is

$$\frac{d\sigma_1}{\sigma_1} = \frac{d\sigma_2}{\sigma_2} = -d\varepsilon_3.$$ (9.15)

Keeping in mind that $d\varepsilon_1/d\varepsilon_2 = (\partial f/\partial \sigma_1)/(\partial f/\partial \sigma_2)$ and the flow rule $d\varepsilon_{ij}$

$= \lambda \, \partial f / \partial \sigma_{ij}$, the properties of equal ratios and (9.15), we can write

$$\frac{\dfrac{\partial f}{\partial \sigma_1} \, d\sigma_1 + \dfrac{\partial f}{\partial \sigma_2} \, d\sigma_2}{\sigma_1 \, d\varepsilon_1 + \sigma_2 \, d\varepsilon_2} = \frac{d\sigma_2 \, \dfrac{\partial f}{\partial \sigma_2}}{\dot{\lambda} \, \sigma_2 \, \dfrac{\partial f}{\partial \sigma_2}} = \frac{d\sigma_2}{\dot{\lambda} \, \sigma_2} = -\frac{d\varepsilon_3}{\dot{\lambda}}$$

$$= \frac{d\varepsilon_1 + d\varepsilon_2}{\dot{\lambda}} = f_1 + f_2.$$

Finally, combining (9.14), (9.15) and the above gives

$$F'(W_{\mathrm{p}}) \leqslant \frac{\partial f}{\partial \sigma_1} + \frac{\partial f}{\partial \sigma_2} \tag{9.16}$$

as the condition for localized necking to be possible.

Now as a specific example we make the following assumptions.

(1) The material yields according to the von Mises criterion; therefore f in (9.13) is given by

$$f = \bar{\sigma} = (\sigma_1^2 - \sigma_1 \sigma_2 + \sigma_2^2)^{1/2}. \tag{9.17}$$

(2) The material strain-hardens isotropically according to the power-law

$$\bar{\sigma} = K \bar{\varepsilon}^n.$$

where $\bar{\sigma}$ and $\bar{\varepsilon}$ are the effective stress and strain respectively.

(3) Strain paths are linear, and we can therefore replace strain increments by total strains.

Now from (9.17)

$$\left. \begin{aligned} \frac{\partial f}{\partial \sigma_1} &= \frac{2\sigma_1 - \sigma_2}{2(\sigma_1^2 - \sigma_1 \sigma_2 + \sigma_2^2)^{1/2}}, \\ \frac{\partial f}{\partial \sigma_2} &= \frac{2\sigma_2 - \sigma_1}{2(\sigma_1^2 - \sigma_1 \sigma_2 + \sigma_2^2)^{1/2}}. \end{aligned} \right\} \tag{9.18}$$

If we define $x = \sigma_2/\sigma_1$, where $\sigma_1 > \sigma_2$ ($\sigma_3 = 0$), then the criterion for localized necking (9.16) becomes

$$F'(W_{\mathrm{p}}) \leqslant \frac{1 + x}{2(1 - x + x^2)^{1/2}}. \tag{9.19}$$

From the assumption of power-law hardening and a von Mises material,

$$F'(W_p) = \frac{\partial f}{\partial W_p} = \frac{d\bar{\sigma}}{dW_p} = \frac{d\bar{\sigma}}{\bar{\sigma}\,d\bar{\varepsilon}} = \frac{n}{\bar{\varepsilon}}. \tag{9.20}$$

The von Mises flow rule $d\varepsilon_{ij} = \dot{\lambda}\,\partial\bar{\sigma}/\partial\sigma_{ij}$ together with the plane-stress state give

$$\frac{\varepsilon_2}{\varepsilon_1} = \frac{1 - 2x}{x - 2}$$

for proportional loading. Then the definition of effective strain leads to

$$\left.\begin{aligned}
\varepsilon_1 &= \frac{(2 - x)\bar{\varepsilon}}{2(1 - x + x^2)^{1/2}}, \\[2mm]
\varepsilon_2 &= \frac{(2x - 1)\bar{\varepsilon}}{2(1 - x + x^2)^{1/2}}.
\end{aligned}\right\} \tag{9.21}$$

Substitution of (9.20) and (9.21) into (9.19) gives the final expression for the limit line of localized necking

$$\varepsilon_1 + \varepsilon_2 = n \quad (\varepsilon_2 \leqslant 0). \tag{9.22}$$

9.4.2 Diffuse necking

The biaxial analogues of Swift's diffuse necking at a maximum in the load in tension are (see §§2.1.1 and 7.2.4)

$$\frac{d\sigma_1}{\sigma_1} \leqslant d\varepsilon_1, \quad \frac{d\sigma_2}{\sigma_2} \leqslant d\varepsilon_2. \tag{9.23}$$

Hill (1952) derived the following condition for diffuse necking by substituting (9.23) into (9.14) and recalling the Lévy–von Mises flow rule:

$$F'(W_p) \leqslant \frac{\sigma_1\left(\dfrac{\partial f}{\partial\sigma_1}\right)^2 + \sigma_2\left(\dfrac{\partial f}{\partial\sigma_2}\right)^2}{\sigma_1\dfrac{\partial f}{\partial\sigma_1} + \sigma_2\dfrac{\partial f}{\partial\sigma_2}}. \tag{9.24}$$

In a way similar to that for localized necking for proportional loading, the locus for diffuse necking is found to be

$$\left.\begin{aligned}
\frac{\varepsilon_1}{n} &= \frac{2(2 - x)(1 - x + x^2)}{(2 - x)^2 + x(2x - 1)^2}, \\[2mm]
\frac{\varepsilon_2}{n} &= \frac{2(2x - 1)(1 - x + x^2)}{(2 - x)^2 + x(2x - 1)^2}.
\end{aligned}\right\} \tag{9.25}$$

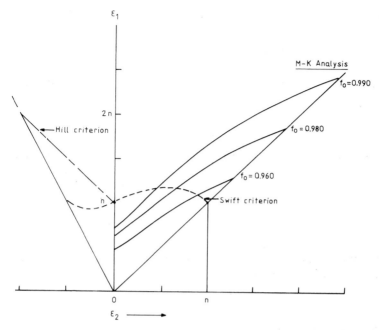

Fig. 9.9 Hill, Swift and Marciniak–Kuczynski necking loci for linear strain paths.

Figure 9.9 shows the loci of limit strains for localized and diffuse necking given by (9.22) and (9.25) respectively.

9.5 The Marciniak imperfection analysis

A bewildering number of different theoretical approaches have been proposed to explain the formation of localized necks in biaxial tensile fields, since we have seen in the previous section that when both in-plane strains are tensile then Hill localized necking is impossible and hence this bifurcation mode is suppressed. In these circumstances another mode of localized necking is observed in which the length of the neck is usually perpendicular to the direction of the major in-plane stress.

To date there have been two broad theoretical frameworks to explain necking in biaxial tensile fields. These are:

(1) the assumption of an initial weakness, imperfection or inhomogeneity in the sheet; as straining proceeds, the imperfection gradually develops into a neck (Marciniak, 1965);

(2) bifurcation analyses describing the initiation of a localized band of straining in an otherwise uniform sheet.

Marciniak and Kuczynski (1967) introduced imperfections into sheets to allow for necking; such an inhomogeneity is shown in Fig. 9.10. In Marciniak and Kuczynski's analysis they assumed that yielding and flow of the material is governed by the von Mises criterion and the Lévy–von Mises equations respectively. Practically, this type of inhomogeneity could be a local thickness variation, which may originate from surface roughness or prior processing, for example orange peeling. As Hutchinson (1979) pointed out, the problem simplifies to a one-dimensional one when the imperfection is a geometrical-thickness variation or a material-property variation that is a function of only the coordinate perpendicular to the infinitely long band. Because of the plane-stress assumption, the stress and strain increments at the minimum part of the neck can be solved directly in terms of the strain increments prescribed outside the band.

Thus in the Marciniak and Kuczynski analysis it is assumed that there is a shallow groove perpendicular to the maximum principal stress, as shown schematically in Fig. 9.10. If we call the groove region B and the area outside the groove region A, we assume that region A is loaded uniformly and is subjected to proportional straining. The initial thickness variation, or inhomogeneity, is given by the thickness ratio

$$f_0 = \left(\frac{t_B}{t_A}\right)_0. \tag{9.26}$$

The stress and strain in region A are given by the conventional effective stress and strain, which can be defined, for example, from the isotropic von Mises

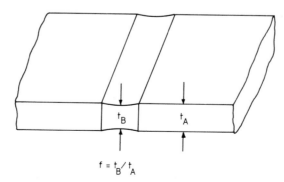

Fig. 9.10 Sheet specimen containing an inhomogeneity in thickness.

yield criterion or Hill's (1948) or (1979) anisotropic yield criteria. In the groove region force equilibrium gives

$$\sigma_{1B} = \sigma_{1A}/f_0; \qquad (9.27)$$

hence σ_{1B} is greater than σ_{1A}. Proportional straining in region A gives

$$\frac{d\varepsilon_{1A}}{d\varepsilon_{2A}} = \frac{\varepsilon_{1A}}{\varepsilon_{2A}} = \text{const.} \qquad (9.28)$$

It is further assumed that the groove strain $d\varepsilon_{2B}$ is always the same as the corresponding strain outside the groove during straining:

$$d\varepsilon_{2A} = d\varepsilon_{2B} = d\varepsilon_2. \qquad (9.29)$$

Although $d\varepsilon_{1A}/d\varepsilon_2$ remains constant during straining, $d\varepsilon_{1B}/d\varepsilon_2$ increases in accordance with (9.27). Hence t_B decreases more quickly than t_A. Now, eventually $d\varepsilon_{1B} \gg d\varepsilon_{1A}$, and deformation in the groove approaches plane strain $d\varepsilon_{2B} = 0$. At this stage ε_{1A}, i.e. the major strain outside the groove, is identified as the limiting value for localized necking. Other interpretations can be used to provide the thinned region. For example, it can be assumed that a metal containing voids can be considered in a similar fashion to a metal with a thinned region. Figure 9.9 shows the valid regions and comparison of Swift, Hill and MK theories. The inhomogeneity theory tends to underestimate the limit strain in the region of $\varepsilon_2 = 0$ and overestimate the strain in the region of balanced biaxial tension. The MK model has been used to study the influence of various material parameters on the forming limit and has been compared with some experimental observations.

The limiting strains are influenced by the yield function in a fundamental way. Figure 9.11(a) shows the effect of r-value on the limit curve for a material with a strain hardening exponent $n = 0.2$ and $f_0 = 0.98$. These curves were calculated using Hill's (1948) anisotropic yield criterion. Now the fact that the level of the curves increases with decreasing r-value is not observed experimentally, particularly for balanced biaxial tension.

For materials with an r-value less than unity, Parmar and Mellor (1978) have used Hill's (1979) anisotropic criterion. They chose an r-value of 0.5, as shown in Fig. 9.11(b). The predicted limit strains vary according to the index of the criterion, and dramatic reductions in the extent of the limit curve can be obtained by taking m-values of 1.7.

Mellor confesses that the choice of f_0 as 0.98 is arbitrary and says that if $f_0 > 0.98$ then the limit strain at balanced biaxial tension will be greater than that obtained experimentally. Mellor's choise of 0.98 for f_0 was probably influenced by Azrin and Backofen's (1970) exhaustive experimental work.

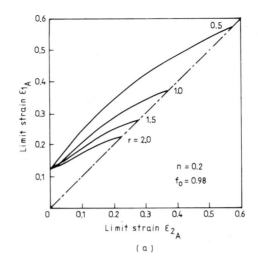

(a)

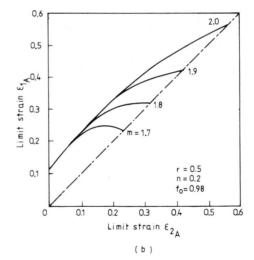

(b)

Fig. 9.11 (a) Theoretical forming-limit curves using Hill's (1948) anisotropic yield criterion. (b) Theoretical forming limit curves using Hill's (1979) nonquadratic yield criterion. (After P. B. Mellor, in *Mechanics of Materials, The Rodney Hill 60th Anniversary Volume* (ed. H. G. Hopkins and M. J. Sewell), pp. 383–415. Pergamon Press, Oxford, 1982.)

Azrin and Backofen subjected a large number of materials to in-plane stretching. They discovered that fractional ratios f_0 of about 0.97 or less were required to obtain agreement between the Marciniak analysis and the experiments. At these levels the grooves should be recognizable with the naked eye—which they were not.

Azrin and Backofen took their experiments one stage further by machining grooves of various depths in sheet specimens. Even with premachined grooves, though, unrealistic groove depths were required before agreement with the Marciniak theory was obtained. To conclude, then, although the Marciniak imperfection analysis may be very convenient to use, there is a gap between the predictions obtained by the MK analysis and experimental data. Hence the study of some other possible mechanisms governing localized necking in biaxial tension is required.

9.6 Theories of bifurcation

9.6.1 Review of bifurcation theories

As mentioned in Chapter 7, another approach to necking is bifurcation, and many of these theoretical approaches to necking have been reviewed by Hutchinson (1979) and Neale (1981). The essential problem with searching for a bifurcation mode under conditions of biaxial tension is the apparent inherent stability of the smooth von Mises yield surface and the Lévy–von Mises flow equations.

Finite-strain versions of the isotropic J_2 deformation theory of plasticity have been proposed by Stören and Rice (1975), Needleman and Tvergaard (1977) and Hutchinson and Neale (1978). Stören and Rice's model assumes rigid plastic deformation, while that of Needleman and Tvergaard allows for possible elastic effects. Essentially Hutchinson and Neale's material is a nonlinear elastic solid that is assumed to be isotropic and incompressible. Neale (1981) extended this to allow for compressibility.

Christofferson and Hutchinson (1979) proposed a J_2 corner theory that coincides with J_2 deformation theory for nearly proportional stress increments. To model the effects of dilitancy in plasticity and pressure sensitive yielding, Rudnicki and Rice (1975) introduced a J_2 flow theory in which isotropic hardening and a smooth yield surface are assumed, but the possibility of a non-normal plastic strain-rate vector is allowed. Gurson (1977) assumed that a deformation mode could be caused by the nucleation and growth of voids (see §5.2.3). Changes in the void volume fraction during plastic deformation are caused by the growth of existing voids, and the nucleation of new cavities by cracking or debonding of inclusions.

9.6.2 Comparison of forming-limit diagrams with theory

Some of the theories for localization and necking in sheet metals have been compared with experiment by Ghosh (1978). Figure 9.12 compares Hill's theory of localized necking and Stören and Rice's theory with punch and in-plane stretching for aluminium-killed steel.

First, it is to be noted that the experiments for the in-plane stretching give values below those for the punch stretching for all strain ratios. In the range for $\varepsilon_2 < 0$ it is shown that Hill's analysis predicts a lower locus than either of the experimental loci for this particular material. Also shown for comparison is the locus derived from the vertex theory of Stören and Rice. Although the Stören–Rice theory predicts a plane-strain limit strain of n, it diverges

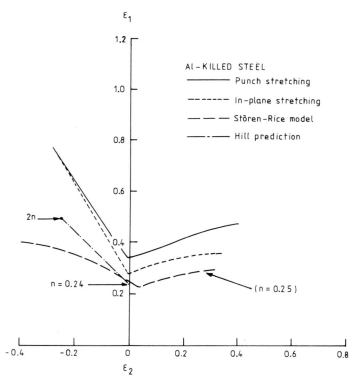

Fig. 9.12 Forming-limit diagram comparing experiments on aluminium-killed steel with predictions from Hill's and Stören and Rice's models. (After A. K. Ghosh, in *Mechanics of Sheet Metal Forming* (ed. D. P. Koistinen and N. M. Wang, pp. 287–312. Plenum Press, New York, 1978.)

markedly from experimental results for $\varepsilon_2 < 0$ and is also significantly lower than the experimental results for $\varepsilon_2 > 0$.

It is quite possible to make modifications to the constitutive equations and indeed increase the complexity of the bifurcation analyses in attempts to obtain closer agreement between theory and experiment as described by Hutchinson (1979) and Neale (1981). The application of bifurcation theory and localization to ductile fracture is clearly a very active area for research, and it is to be hoped that further important discoveries will continue to be made in the theory.

9.7 Improvements in drawing

9.7.1 Limiting drawing ratio in cup drawing

In cup drawing formability is expressed in terms of a limiting drawing ratio, which is the largest ratio of blank to cup diameters that can be drawn without failure. Following Hosford and Caddell (1983), we assume that the plastic flow in the flange occurs in plane strain with $d\varepsilon_z = 0$; this means that there is no change in thickness, and if it is assumed that the surface area remains constant, then for an element at radius ρ from the centre (Fig. 9.13)

$$\pi\rho^2 + 2\pi r_1 h = \pi\rho_0^2. \tag{9.30}$$

Then $d\rho = -r_1\,dh/\rho$ and, because the circumference of the element is proportional to ρ, $d\varepsilon_y = d\rho/\rho$. Because of volume constancy,

$$d\varepsilon_x = -d\varepsilon_y = -\frac{d\rho}{\rho}. \tag{9.31}$$

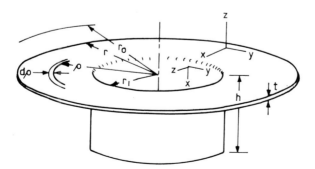

Fig. 9.13 A partially drawn cup.

The incremental work done on the flange element at ρ with increment $d\rho$ is given by

$$dW = 2\pi t\rho \, d\rho \, (\sigma_x \, d\varepsilon_x + \sigma_y \, d\varepsilon_y)$$

$$= \frac{2\pi t\rho \, d\rho \, (\sigma_x - \sigma_y) r_1 \, dh}{\rho^2} \tag{9.32}$$

after substituting (9.31) and $d\rho = -r_1 \, dh/\rho$ into (9.32), The values of σ_x and σ_y will vary with position, but it is reasonable to assume that $\sigma_x - \sigma_y$ is constant. We shall call this stress difference the flange flow strength σ_F. The total work on all elements per increment of punch travel dh is

$$\frac{dW}{dh} = \int_{r_1}^{r} \frac{2\pi r_1 t \sigma_F \, d\rho}{\rho} = 2\pi r_1 t \sigma_F \ln\left(\frac{r}{r_1}\right). \tag{9.33}$$

The drawing force F_d will be largest at the beginning of drawing, $r = r_0$, so

$$F_{d\,max} = 2\pi r_1 t \sigma_F \ln\left(\frac{r_0}{r_1}\right) = 2\pi r_1 t \sigma_F \ln\left(\frac{d_0}{d_1}\right), \tag{9.34}$$

where d_0 and d_1 are the blank and cup diameters respectively. The cup wall must carry the stress $F_d/2\pi r_1 t$. The maximum wall stress is given by

$$\sigma_x = \frac{F_{d\,max}}{2\pi r_1 t} = \sigma_F \ln\left(\frac{d_0}{d_1}\right). \tag{9.35}$$

The drawing limit will be reached when the flow strength in the walls σ_w is reached:

$$\sigma_w = \sigma_F \ln\left(\frac{r_0}{r_1}\right) \tag{9.36}$$

and therefore

$$LDR = \left(\frac{d_0}{d_1}\right)_{max} = \exp\left(\frac{\sigma_w}{\sigma_F}\right). \tag{9.37}$$

For an isotropic material $\sigma_w = \sigma_F$, and therefore $LDR = e$. In practice though, the LDR is usually closer to 2. This discrepancy is caused by neglecting friction and bending in the above analysis.

It is clear that anisotropy should affect the LDR. If Hill's (1948) anisotropic criterion is used and we assume planar isotropy then it can be shown that

$$\ln\left(\frac{d_0}{d_1}\right)_{max} = \left(\frac{1+r}{2}\right)^{1/2}, \tag{9.38}$$

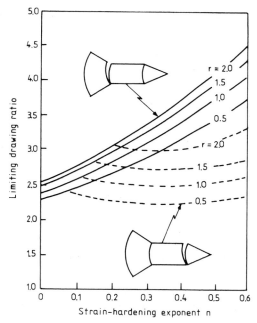

Fig. 9.14   Effect of r- and n-values on LDR and position of fracture. (After M. G. El-Sebaie and P. B. Mellor, *Int. J. Mech. Sci.* **15**, 485–501 (1973).)

where r is the strain ratio. It is possible to introduce an efficiency factor η, which will vary with lubrication, hold-down force, sheet thickness and die radius; thus

$$\ln\left(\frac{d_0}{d_1}\right)_{\text{max}} = \eta\left(\frac{1+r}{2}\right)^{1/2}. \tag{9.39}$$

Strain hardening can be incorporated in the above analysis. Figure 9.14 shows the effect of r- and n-values on the LDR. The type of failure is independent of n-value. El-Sebaie and Mellor (1972, 1973) showed that under certain conditions the failure mode changes to necking under simple tension in material emerging from the flange; this failure mode is very sensitive to n-value, but not so sensitive to r-value. Ng *et al.* (1976), among many other researchers, showed how different conditions of lubrication can affect the LDR. The effect of r-values less than unity on the LDR was investigated by Dodd and Atkins (1983b) using case (iv) of Hill's (1979) criterion. These researchers found that using this yield criterion has little effect on the predicted LDR.

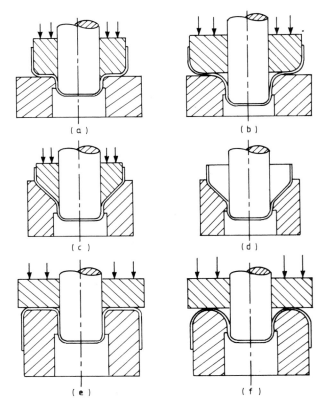

Fig. 9.15 Some possible methods of redrawing; (a)–(d) direct and (e) and (f) reverse redrawing. (Reproduced with permission from W. Johnson and P. B. Mellor *Engineering Plasticity*, published by Ellis Horwood Ltd, Chichester, England, 1983.)

9.7.2 Redrawing and ironing

Sometimes cups with larger height-to-diameter ratios are required than can be produced by a single draw. In these cases redrawing and/or ironing is carried out. There are two categories of redrawing, as described by Johnson and Mellor (1983): direct redrawing and reverse redrawing. In reverse redrawing the cup is turned inside out. Some possible redrawing processes are shown in Fig. 9.15.

Wall thinning by ironing is used to produce a more uniform thickness along a cup wall, and the results of the process do not improve with increasing *r*-value. Figure 9.16 shows a combined drawing, redrawing and ironing sequence utilizing concentric punches.

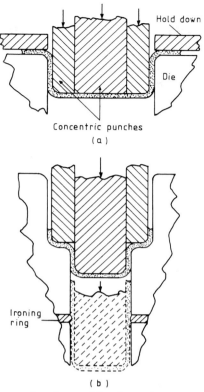

Fig. 9.16 Deep drawing, redrawing and ironing using concentric punches. (From W. F. Hosford and R. M. Caddell, METAL FORMING: Mechanics and Metallurgy, © 1983, p. 291. Reprinted by permission of Prentice-Hall, Englewood Cliffs, New Jersey.)

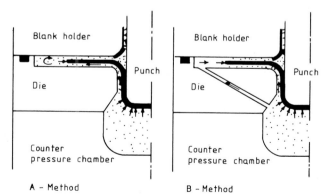

Fig. 9.17 Two methods of counter pressure deep drawing. (After K. Nakamura and T. Nakagawa, Preprint 427, Spring Conf. on Plastic Working, 1983 (in Japanese).)

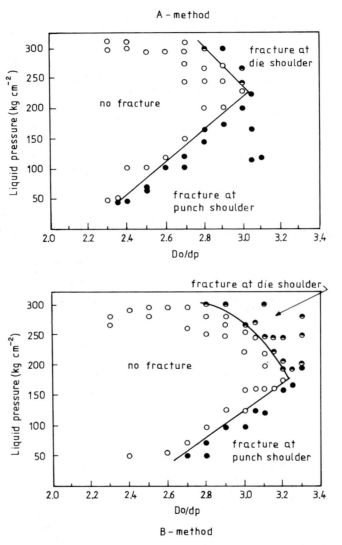

Fig. 9.18 Forming limits for the two methods of counter-pressure drawing, showing peak LDRs at intermediate liquid pressures. (After K. Nakamura and T. Nakagawa, Preprint 427, Spring Conf. on Plastic Working, 1983 (in Japanese).)

9.7.3   Drawing in the presence of a superimposed hydrostatic pressure

Two methods of deep drawing against a hydrostatic pressure were proposed by Nakamura and Nakagawa (1982). The drawing methods are shown in Fig. 9.17. Nakamura and Nakagawa refer to these techniques of drawing as counter-liquid-pressure (CLP) deep drawing. The pressurizing liquid not only provides a counter pressure but also lubricates the contacting surfaces between the die, the blank holder and the blank. Also the liquid decreases susceptibility to flange wrinkling. Method B should be more successful than method A because it is possible to control the pressure more accurately.

The forming limits of both methods are shown in Fig. 9.18 for annealed commercially pure aluminium. The highest LDRs occur at intermediate values of the liquid pressure. Typical maximum LDRs are about 3 for method A and about 3.2 for method B.

The CLP concept has also been applied to reverse redrawing by Nakamura and Nakagawa (1983). The processing sequence is shown in Fig. 9.19. The LDRs after the first and final stages are shown in Fig. 9.20; once

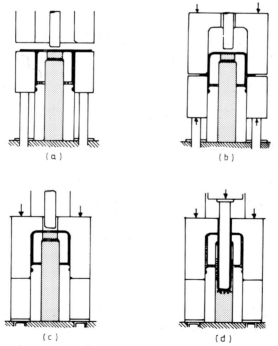

Fig. 9.19   Processing sequence for reverse redrawing with a counter pressure. (After K. Nakamura and T. Nakagawa, Preprint 427, Spring Conf. on Plastic Working, 1983 (in Japanese).)

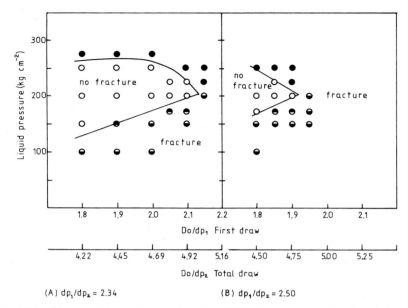

Fig. 9.20 First- and second-stage drawing LDRs in reverse redrawing. (After K. Nakamura and T. Nakagawa, Preprint 427, Spring Conf. on Plastic Working, 1983 (in Japanese).)

again, intermediate pressures produce redrawn cups with no fractures. It is possible to produce cups with final LDRs as high as 4.7 using this technique, (see also Nakamura and Nakagawa, 1984). This research of Nakagawa and coworkers illustrates the importance of utilizing the transfer of the drawing load to the punch so as to reduce the load on the lower cup wall and the fundamental effect of hydrostatic pressure on metal ductility. Clearly, conceptually counter pressures can be used to advantage in other sheet-metal working processes.

A novel method for improving the LDR in steel cupping is to utilize the advantages of local quenching (Machida and Nakagawa, 1978). In this technique the material constituting the cup wall or the material in the region of the punch profile is rapidly heated in an induction furnace and then quenched to increase its resistance to stretching. A low-carbon steel was used so that the material did not become very brittle after quenching. Improvements in the LDR from a conventional value of 2.07 up to 2.51 were obtained by adjusting heating times and temperatures.

10

Workability in
Bulk Forming Processes

Three classes of fracture or cracking occur in bulk forming operations: (i) free-surface cracks; (ii) cracks originating from the die–workpiece interfaces; (iii) internal cracks, sometimes referred to as centre bursts. Free-surface cracks occur, for example, at the edges of rolled strip and at the surfaces of upset or forged components; recent reviews of surface cracking in these processes have been given by Dodd and Boddington (1980) and Jenner and Dodd (1981). The straight-line fracture condition for surface cracking in axisymmetrical upsetting obtained by Kudo and Aoi (1967) has been described in Chapter 8 and is shown in Fig. 8.13. More recent work by Kuhn and his coworkers has confirmed how valuable this fracture condition can be in industrial processing. For example, the work of Ertürk (1979) shows the effects of second-phase particle shape and heat treatment on the forming-limit line. Figure 10.1(a) shows the effect of ageing on a 2014 aluminium alloy; the overaged structure has the highest workability, whereas the solution-treated structure has the lowest workability. Figures 10.1(b–d) all show that structures of steels with coarse spheroidal carbides have the greatest workability.

Lee (1972) and others have shown that the onset of surface cracking in bending and rolling coincide with the straight-line fracture condition, and therefore the fracture limit has a wide validity. Die–workpiece interface fractures have been observed by Jain and Kobayashi (1970), who carried out plane-strain side pressings of 7075 aluminium alloy with and without lubrication. The fracture initiation points and propagation planes were compared with the predicted flow patterns of slip-line field analyses. Typical slip-line fields are shown in Fig. 10.2. Fairly close correlations were observed between experiment and theory.

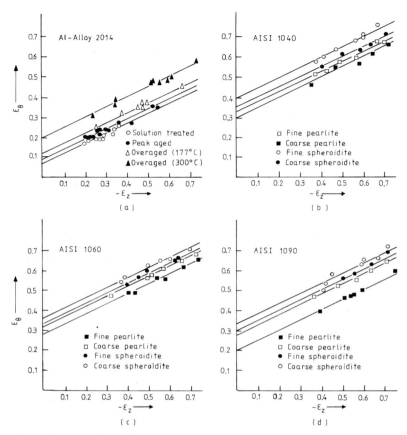

Fig. 10.1 Forming limits in upsetting of an aluminium alloy and three steels. (a) Extensive overaging improves formability. (b)–(d). Heat treatment to produce coarse pearlite and coarse spheroids raises the forming-limit lines markedly. (After T. Ertürk, in *Proc. Int. Conf. on the Mechanical Properties of Materials 3*, Vol. 2 (ed. K. J. Miller and R. F. Smith), pp. 653–662 (1979).)

Intense localized shear flow was revealed by etching, and in most cases fracture occurred, usually along a velocity discontinuity in the slip-line field.

More recently, Clift *et al.* (1985) found similar results for aluminium-alloy specimens with machined flats. The work of both Jain and Kobayashi and Clift *et al.* clearly shows the importance of the interaction of die, workpiece and friction in the generation of die–workpiece interface fractures.

Internal cracks are usually associated with rod extrusion and wire drawing. For certain combinations of die angle, draft and friction, a significant

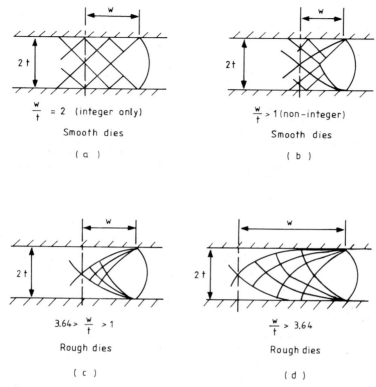

Fig. 10.2 Slip-line fields for side pressing between smooth dies (a, b) and rough dies (c, d). Note that for integer values of w/t the slip-lines are straight as in (a).

hydrostatic tension can be included at the centreline of the component. This centreline hydrostatic tension, together with any centreline segregation or porosity, can induce void nucleation and growth. These aspects of sheet drawing have been studied at great depth by Rogers and Coffin (1971), who confirmed by slip-line field analysis, combined with detailed experiments, that centre bursting occurred when the midplane hydrostatic tension became greater than the plane-strain yield stress.

10.1 Forging

This is one of the oldest types of mechanical forming process, in which a workpiece, usually hot, is plastically compressed by hammering. Often very large reductions are achieved, for example, hand hammering of a hot steel workpiece can easily result in reductions in thickness of 90%.

There are two broad categories of forging process; these are open- and closed-die (impression-die) forging. In open-die forging flat dies are usually used, and it is primarily the frictional constraint at the die–workpiece interface that determines the spread of the metal during reduction. On the other hand, in closed-die forging the dies themselves are used to shape the metal to a near-finished shape.

Hot open-die forging is often used as the first stage in the successive breakdown of the structure of ingots. As ingots are normally quite large, this type of forging is carried out progressively along the length of the workpiece. The object of ingot forging, called cogging, is to transform the structure from that of a weak casting with its columnar grains and pipe to a uniform equiaxed microstructure.

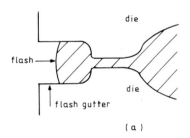

(a)

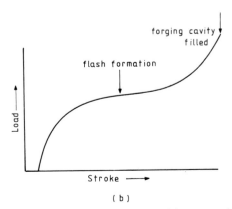

(b)

Fig. 10.3 (a) A flash gutter into which excess metal has spread. (b) Typical load–stroke diagram for closed-die forging showing that the flash begins to form before the forging cavity is completely filled.

Cold open-die forging is normally used as a secondary process such as upsetting and heading, which have already been described in Chapter 8. In these processes tensile stresses at the expanding free surface of the workpiece are induced, as we have seen. Clearly, frictional constraint, reduction and microstructure are initimately related to the formability of the workpiece.

Closed-die or impression-die forging is used to manufacture components to fairly tight tolerances. In this class of process the workpiece is roughly preformed in one set of dies; this produces a forging, approaching the final shape. The forging is next accurately forged to its final shape by the finishing dies. To ensure that there is enough metal in the forging, slightly more metal than necessary is used. Any excess metal then spreads out of the die cavity into the flash gutter, as shown in Fig. 10.3(a). The excess metal is called the flash. A typical load–stroke diagram is shown in Fig. 10.3(b). The flash begins to form before the die cavity is filled completely, and it is only at relatively large loads that the forging is formed completely. Flash-gutter design is very important because the formation of the flash determines the escape rate of metal from the cavity, which in turn determines the flow of the metal in the part. An excellent introduction to open- and closed-die forging is given by Dieter (1976).

10.2 Application of ductile fracture theories to forging

10.2.1 Theories of free-surface ductility

The fracture-limit locus for surface cracking approximates to a straight line, as shown in Fig. 8.13 for a medium-carbon steel. Several approaches have been proposed to fit the line and explain the fracture patterns observed.

For convenience, we now give some basic expressions for stresses and strains on an equatorial surface element ($\sigma_r = 0$):

$$d\bar{\varepsilon} = \tfrac{1}{3}[2\{(d\varepsilon_\theta - d\varepsilon_z)^2 + (d\varepsilon_z - d\varepsilon_r)^2 + (d\varepsilon_r - d\varepsilon_\theta)^2\}^{1/2}]$$

$$= \frac{2}{\sqrt{3}}(\alpha^2 + \alpha + 1)^{1/2}\,d\varepsilon_z \tag{10.1}$$

where $\alpha = d\varepsilon_\theta/d\varepsilon_z$, and using the incompressibility condition $-d\varepsilon_r = d\varepsilon_z + d\varepsilon_\theta$, we have

$$\bar{\sigma} = [\tfrac{1}{2}\{(\sigma_\theta - \sigma_z)^2 + (\sigma_z - \sigma_r)^2 + (\sigma_r - \sigma_\theta)^2\}]^{1/2} = \sigma_z(x^2 - x + 1)^{1/2}$$

$$= \frac{\sigma_z\sqrt{3}(\alpha^2 + \alpha + 1)^{1/2}}{\alpha + 2}, \tag{10.2}$$

where $x = \sigma_\theta/\sigma_z$,

$$\alpha = \frac{2x - 1}{2 - x}, \qquad x = \frac{2\alpha + 1}{\alpha + 2}. \tag{10.3}$$

Using flow theory, $d\varepsilon_{ij} = \dot{\lambda}(\sigma_{ij} - \sigma_m)$.

The Latham criterion (8.4), which states that fracture will occur when the tensile plastic work reaches some critical value, has been reformulated by Lee (1972)—see also Kuhn (1978)—in terms of the local hoop and axial strains at the equator of a cylinder, ε_θ and ε_z respectively.

It is worth noting that from (10.3) the strain path $\alpha = -\frac{1}{2}$, i.e. $d\varepsilon_\theta = d\varepsilon_r = -\frac{1}{2} d\varepsilon_z$, leads to $x = 0$, that is a uniaxial uniform compression. α remains negative and less than $-\frac{1}{2}$; otherwise σ_θ would be compressive during forging. This is supported by experimental observations, and leads to the hoop stress σ_θ being the maximum tensile stress for any strain path ($\alpha < -\frac{1}{2}$); see Fig. 10.4. From (10.2) and (10.3),

$$\bar{\sigma} = \frac{\sqrt{3}\,(\alpha^2 + \alpha + 1)^{1/2}\,\sigma_\theta}{2\alpha + 1}. \tag{10.4}$$

Latham's criterion can be deduced as follows (the assumptions involved are

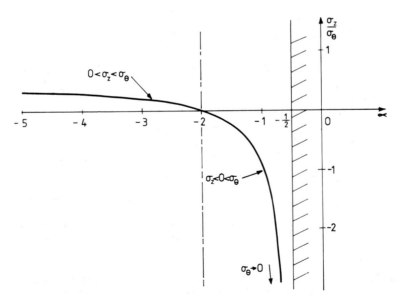

Fig. 10.4   Limitations on σ_z and σ_θ.

described in parentheses):

$$C = \int_0^{\bar{\varepsilon}_f} \sigma_T \, d\bar{\varepsilon} = \int_0^{\bar{\varepsilon}_f} \sigma_T \, d\bar{\varepsilon} = \int_0^{\bar{\varepsilon}_f} \frac{2\alpha + 1}{\sqrt{3}\,(\alpha^2 + \alpha + 1)^{1/2}} \bar{\sigma} \, d\bar{\varepsilon}$$

$$= \frac{2\alpha + 1}{\sqrt{3}\,(\alpha^2 + \alpha + 1)^{1/2}} \int_0^{\bar{\varepsilon}_f} \bar{\sigma} \, d\bar{\varepsilon} \qquad \text{(for linear strain paths)}$$

$$= \frac{2\alpha + 1}{\sqrt{3}\,(\alpha^2 + \alpha + 1)^{1/2}} K\bar{\varepsilon}_f \qquad \text{(power-law hardening, } \bar{\sigma} = K\bar{\varepsilon}^n\text{,}$$

$$n = 0; \text{ hence } \bar{\sigma} = K)$$

$$= 2K(2\alpha + 1)\varepsilon_{zf}$$

$$= \tfrac{2}{3}K(2\sigma_{\theta f} + \sigma_{zf}). \tag{10.5}$$

It is found that the criterion gives an approximately linear relationship for small strain-hardening exponents, as shown in Fig. 10.5. Importantly, it coincides with the linear fracture condition when the strain-hardening exponent is zero. However, Latham's criterion cannot predict the difference in fracture modes as a function of stress state. Moreover, (8.4) is empirical.

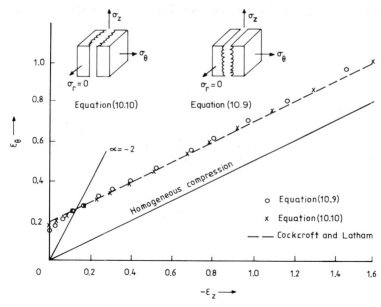

Fig. 10.5 Latham and McClintock's fracture criteria applied to free-surface ductility. There are two possible fracture planes, which one is activated depends on the value of the strain ratio α.

Oh and Kobayashi (1976a) showed that the Oyane model for ductile fracture, equation (5.36) based on porous plasticity, gives another fracture locus. The deduction is as follows. If

$$C = \int_0^{\bar{\varepsilon}_f} \left(a_0 + \frac{\sigma_m}{\bar{\sigma}} \right) d\bar{\varepsilon}$$

then

$$C = a_0 \bar{\varepsilon}_f + \int_0^{\bar{\varepsilon}_f} \frac{\sigma_m}{\bar{\sigma}} \, d\bar{\varepsilon}$$

$$= a_0 \bar{\varepsilon}_f + \int_0^{\bar{\varepsilon}_f} \frac{\alpha + 1}{\sqrt{3} \, (\alpha^2 + \alpha + 1)^{1/2}} \, d\bar{\varepsilon}$$

$$= a_0 \bar{\varepsilon}_f + \frac{\alpha + 1}{\sqrt{3} \, (\alpha^2 + \alpha + 1)^{1/2}} \, \bar{\varepsilon}_f$$

(for linear strain paths)

and

$$a_0 \frac{2}{\sqrt{3}} (\varepsilon_{zf}^2 + \varepsilon_{zf}\varepsilon_{\theta f} + \varepsilon_{\theta f}^2)^{1/2} + \tfrac{2}{3}(\varepsilon_{zf} + \varepsilon_{\theta f}). \tag{10.6}$$

The advantage of (10.6) is that it utilizes porous plasticity theory as its basis and only the assumption of a linear strain path is used in its deduction. But the expression is nonlinear. Even in the case of $\varepsilon_\theta \gg |\varepsilon_z|$ or $\alpha \to -\infty$ (10.6) becomes

$$a_0 \frac{2}{\sqrt{3}} (\varepsilon_{\theta f} + \tfrac{1}{2}\varepsilon_{zf}) + \tfrac{2}{3}(\varepsilon_{\theta f} + \varepsilon_{zf}) = C, \tag{10.7}$$

if we can disregard ratios of $\varepsilon_z/\varepsilon_\theta$ with powers greater than unity. Equation (10.7) is linear, but the slope cannot be $-\tfrac{1}{2}$; it is between -1 and $-\tfrac{1}{2}$. Like the Latham criterion this criterion does not explain the different fracture modes observed experimentally.

Oh et al. (1979) and Oh and Kobayashi (1976a) applied the McClintock analysis of hole growth to fracture at the equatorial free surface of an expanding cylinder. Starting from the McClintock equation for fracture given by (5.21),

$$\bar{\varepsilon}_f = \frac{\ln(l_b^0/2b_0)}{\left[\dfrac{\sqrt{3}}{2(1-n)} \sinh \left\{ \dfrac{\sqrt{3}(1-n)\,(\sigma_a + \sigma_b)}{2} \dfrac{}{\bar{\sigma}} \right\} \right]} \tag{10.8}$$

the theory is used to describe surface cracking in upsetting. Following

Kobayashi *et al.*, there are two possible modes of fracture. However, in the case under consideration, σ_θ is always the maximum tensile stress; then $\sigma_a = \sigma_\theta$. (Taking $\ln(l_b^0/2b_0)$ to be a material constant, assuming a vanishingly small strain-hardening exponent and recalling (10.1) the following expressions are obtained:

$$\varepsilon_{\theta f} = \frac{\alpha}{2}\left\{\frac{3}{1+\alpha+\alpha^2}\right\}^{1/2} \cdot \frac{K}{3^{1/2}\sinh\left\{\dfrac{3^{1/2}(\sigma_\theta + \sigma_2)}{2\bar\sigma}\right\}\Big/2} \qquad (10.9)$$

for a plane-strain stress state on the (θ, z) planes; and

$$\varepsilon_{\theta f} = \frac{\alpha}{2}\left\{\frac{3}{1+\alpha+\alpha^2}\right\}^{1/2} \cdot \frac{K}{3^{1/2}\sinh\left(\dfrac{3^{1/2}\sigma_\theta}{2\bar\sigma}\right)\Big/2} \qquad (10.10)$$

for a plane-strain stress state on the (θ, r) planes. The results of these two equations are shown in Fig. 10.5. Both equations predict an approximately linear fracture locus. It is important to note, however, that although two modes of void growth are presumed, the theory only predicts one fracture plane normal to the maximum tensile stress. Unfortunately, this predicted fracture plane does not coincide with either of the two planes observed in experiments. The above predictions are all based on void-growth theories or related empirical criteria.

Among the most important experimental results in upsetting and related processes are those of Lee (1972); these experiments have been mentioned in Chapter 8. Using very fine grids on the equatorial free surfaces of three medium-carbon steels, Lee measured strains to fracture in upsetting, bending and rolling. His key observation was that there was a perturbation in the local axial strain, corresponding to $d\varepsilon_z = 0$. From detailed metallographic observations, Lee concluded that extensive microstructural damage occurred towards the end of the strain perturbation, i.e. rapid void nucleation and growth took place. From these experimental findings, it can be concluded that the free-surface forming limit in upsetting and related processes coincides with a transition to a plane-strain stress $d\varepsilon_\theta \neq 0$, $d\varepsilon_r \neq 0$, $d\varepsilon_z = 0$, $\sigma_\theta \neq 0$, $\sigma_z \neq 0$ and $\sigma_r = 0$.

The idea of localized necking or at least a plane-strain stress state at the equatorial free surface of an expanding cylinder led Lee to propose modifications to the Marciniak–Kuczynski (1967) model, used originally to predict forming limits in sheet forming. Lee developed two models, one with an incipient neck along the r-axis and the other with a neck along the z-axis. The theoretical fracture loci following from these two models are shown in Fig. 10.6. As in sheet-forming problems, the levels of the fracture loci are very

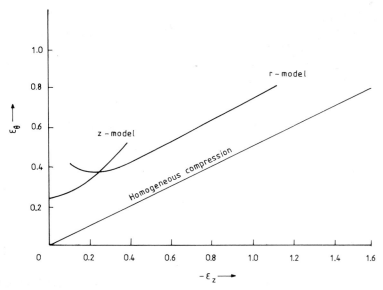

Fig. 10.6   The two Marciniak necking models plotted on the forming-limit diagram.

sensitive to the inhomogeneity factor. One serious doubt still remains about the application of this imperfection model to upsetting, namely the difference in boundary conditions. In the case of an element of a sheet it is physically reasonable that a neck may develop that is symmetrically disposed on both sides of the element. However, the same type of behaviour is apparently impossible in upsetting because an element only has one free surface, and it will therefore be constrained from necking. Jenner *et al.* (1981) proposed another approach to surface cracking in upsetting based on shear localiza-tion, and Bai and Dodd (1985) explained the shear fracture line as well as different fracture modes. The criterion for instability in simple shear is given by a critical shear strain γ_i, for instance a critical adiabatic strain. To apply the criterion to upsetting, it is necessary to consider shear deformation on an infinitesimally small cubical element at the equatorial free surface of a cylinder. Although the circumstances are not those of pure shear in the element, the orientations of the cracks have a close connection with the bands of shear localization on the maximum-shear planes. Bai and Dodd assume that shear localization can occur in an element when (1) a simple shear-instability condition is satisfied, and (2) at that moment there is no extensional increment along the maximum-shear plane under consideration, i.e. either $d\varepsilon_\theta$, $d\varepsilon_r$ or $d\varepsilon_z = 0$. There are three possible pairs of planes upon which shear instability can occur: (i) the planes oriented at 45° to both the θ- and r-axes (mode I); (ii) the planes oriented at 45° to both the θ- and z-axes (mode II); (iii) the planes oriented at 45° to the z- and r-axes (mode III). Which of these planes undergoes instability will depend on the stress and strain histories. In terms of the $(\varepsilon_\theta, \varepsilon_z)$ plane these three fracture modes are

$$
\left.
\begin{aligned}
\varepsilon_\theta &= \tfrac{1}{2}\gamma_i - \tfrac{1}{2}\varepsilon_z && \text{for mode I,} \\
\varepsilon_\theta &= \gamma_i + \varepsilon_z && \text{for mode II,} \\
\varepsilon_\theta &= \gamma_i \pm 2\varepsilon_z && \text{for mode III.}
\end{aligned}
\right\}
\tag{10.11}
$$

These possibilities are shown in Fig. 10.7. Mode III instability will be absent in conventional upsetting because of the impossibility of meeting the requirement $\alpha = 0$ at instability. From the tests by Kudo and Aoi, all the equatorial strain paths in upsetting follow

$$
\alpha = \frac{\mathrm{d}\varepsilon_\theta}{\mathrm{d}\varepsilon_z} < -\frac{1}{2} \quad \text{and} \quad \frac{\mathrm{d}^2\varepsilon_\theta}{\mathrm{d}|\varepsilon_z|^2} \geq 0.
\tag{10.12}
$$

Hence α decreases monotonically with height reduction. Therefore, once α reaches $-\infty$, which decides mode I instability, the element can never "escape" from that state. This leads to mode I being absolutely unstable.

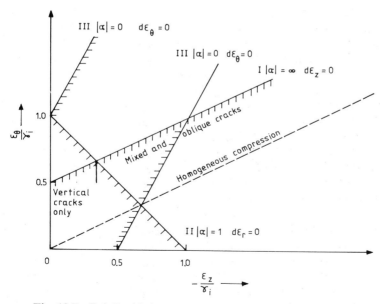

Fig. 10.7   Bai–Dodd theory applied to free-surface cracking.

However, although $\alpha = -1$ and (10.11) define mode II instability, it is possible that α can drop below -1 with further reduction and then violate one of the requirements of instability. Therefore mode II becomes metastable, with excess shear strain on the corresponding shear plane, until the strain path meets the mode I instability condition. Although this theory can explain both the linear fracture limit and different fracture modes, several empirical assumptions are incorporated into it. Hence further verification is needed.

10.2.2   Upper-bound technique applied to internal cracking

Internal cracks sometimes occur in forgings. Causes can be high friction at the dies or the die shape itself, which can restrict metal flow; examples are shown in Fig. 10.8.

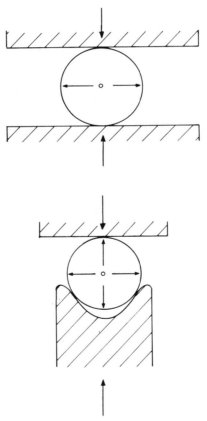

Fig. 10.8 Tensile stresses induced at the central region of workpieces by different shaped tools.

Both Avitzur (1977) and Kudo (1960) applied the upper-bound technique to the onset of cavity formation. Using this technique, the workpiece is divided into a number of rigid blocks. As a result of the externally applied load, shear deformation is assumed to occur between the rigid blocks along velocity discontinuities. The power dissipated by shearing in the workpiece is equated to the power applied to the tooling. If a kinematically admissible velocity field is chosen then the equating of the power terms will lead to an upper bound to the required load. The upper-bound theorem gives

$$\Delta v = \dot{U}\left\{1 + \left(\frac{w}{t}\right)^2\right\}^{1/2}. \tag{10.13}$$

where S_v is the length of the velocity discontinuities, S_T is the length over

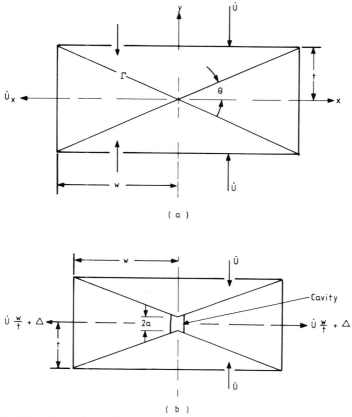

(a)

(b)

Fig. 10.9 Upper-bound solutions for plane-strain compression (a) without and (b) with a central cavity.

which the surface tractions T_i are applied, τ is the shear stress and Δv is the magnitude of the discontinuities.

Let us take as our example the plane-strain compression of a rectangular block, after Avitzur (1977) and Kudo (1960), as shown in Fig. 10.9(a). Each block moves as a rigid body, but shearing occurs along the lines of discontinuity. The velocity discontinuity along a line is given by

$$\Delta v = \dot{U}\left\{1 + \left(\frac{w}{t}\right)^2\right\}^{1/2}. \tag{10.14}$$

Here $\dot{U}$ is the velocity of the dies, $2w$ is the width of die workpiece contact and $2t$ is the thickness of the workpiece. If central cracking is a possibility then we may modify the velocity field as shown in Fig. 10.9(b). Here there is a central cavity with a height of $2a$. Because of the presence of the cavity, the outward motion of the rigid blocks on either side of the cavity is faster than that shown in Fig. 10.9(a), and the velocity discontinuity along the surface is

$$\Delta v = \left[1 + \left\{\frac{w}{t-a}\right\}^2\right]^{1/2}\dot{U}. \tag{10.15}$$

If a approaches zero then Δv is given by (10.14). Avitzur also allows for a further term in the upper-bound theorem statement that corresponds to a rate of work of pore formation and growth $\dot{W}_{\text{pore}}$. So the external work rate W_{EX} is equated to the internal rate of shearing plus the rate of pore formation:

$$\dot{W}_{\text{EX}} = \dot{W}_{\text{IN}} + \dot{W}_{\text{pore}}.$$

The concept of cavities together with triangular rigid elements is not new, and was originally discussed by Kudo (1960).

10.3 Rolling and related cracking

Rolling is the reduction in thickness of a metal by the squeezing action of two rotating rolls. There are two classes of rolling process: flat rolling and section rolling. In flat rolling thin strip is produced from plate or ingot, whereas in section rolling complex sections such as I-beams and railroad rails are produced. For strip rolling the rolls are always cambered a little so that the elastic deflection of the rolls resulting from rolling will produce a nearly parallel roll gap. A strip should therefore have no gauge variation, but in practice it is manufactured with a slight crown, which stops lateral wandering in the rolls. Lack of flatness manifests itself in four ways. The four defects that can be produced are shown in Fig. 10.10. Loose edge and loose middle are the result of mismatching between the roll gap and the strip. Herringbone and

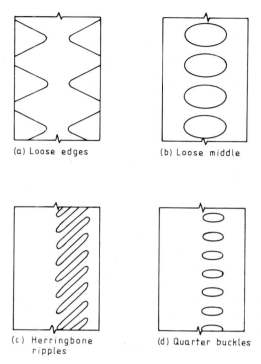

(a) Loose edges   (b) Loose middle

(c) Herringbone   (d) Quarter buckles
    ripples

Fig. 10.10   The four most common shape defects in strip.

quarter buckles are caused by locally excessive reductions. Under severe conditions loose edges and loose middle can result in edge cracks or zipper cracks.

10.3.1   Edge cracking

Edge cracking in rolling was first studied systematically by Hessenberg and Bourne (1949) and Bourne (1951), and the causes have been reviewed by Dodd and Boddington (1980). Important variables that affect the onset of edge cracking are reduction per pass, frictional conditions, material spread, edge shape and mechanical properties of the work material. Spread is defined as the extension of material along the roll axes, and is clearly of importance when rolling sections of small width–thickness ratio and large drafts per pass. It has been confirmed by a number of researchers that initial edge shape is one of the most important factors in determining the onset of edge cracking. Initially square-edged strip has a significantly greater reduction to fracture than do initial radiused or chamfered edges. It was found by Lee (1972) and others that the strain paths to fracture for edge cracking in rolling do terminate on the same straight line as points for surface cracking in upsetting. Therefore it follows that the linear fracture condition is valuable when extended to edge cracking in rolling. Two obvious ways of overcoming the danger of edge cracking is to induce compressive stress at the surface by suitably shaped rolls or (more wastefully) periodically slitting the rounded edges between rolling passes.

10.3.2 Propagation of cracks

Oh and Kobayashi (1976b) have shown the effect of lubrication on the rolling of 7075 aluminium alloy with initially square edges. For unlubricated rolling, cracks, if any, are confined to the edges. However, for lubricated rolling the cracks propagate catastrophically across the entire widths of the specimens, as shown in Fig. 10.11.

Usually, as soon as the onset of edge cracking occurs then this is taken to be the workability limit in rolling. However, despite this, Dodd and Boddington (1980) rolled aluminium alloy strips with initial width–thickness ratios of 4 ($w \approx 25$ mm) with initially square and radiused edges. After edge cracking occurred, the strips were further reduced in thickness, the cracks curved round and eventually grew parallel to the rolling direction, as shown in Fig. 10.12. For given rolling conditions the width of the uncracked

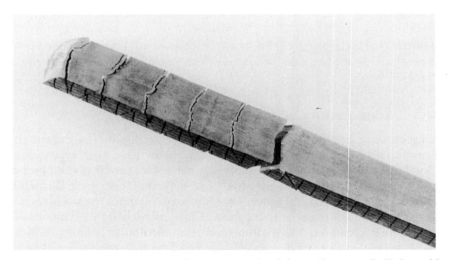

Fig. 10.11 Typical cracks in rolling in 7075 aluminium. (Courtesy S. Kobayashi, after S. I. Oh and S. Kobayashi, AAFML-TR-76-61 (1976).)

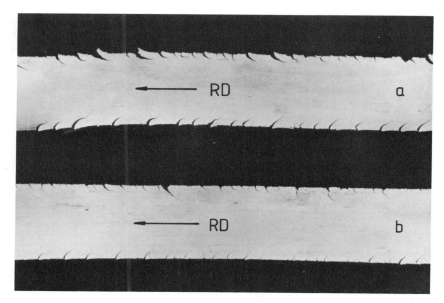

Fig. 10.12 Directions of crack propagation in an aluminium alloy rolled to the minimum thickness obtainable on the mill (~0.24 mm): (a) 6.35 mm initial edge radius; (b) initially square edge, initial width 25 mm. (After B. Dodd and P. Boddington, *J. Mech. Working Techn.* **3**, 239–252 (1980).)

material at the minimum thickness was insensitive to initial edge shape. The width of the central uncracked material is a measure of the region of the material subjected to plane-strain conditions. From these experiments it can be concluded that the onset of edge cracking in rolling need not signal the workability limit if a large fraction of the width remains uncracked. However, with relatively brittle materials the onset of edge cracking is necessarily the workability limit.

10.3.3 Alligatoring

Alligatoring or crocodile cracking is a form of catastrophic midplane splitting that occurs in rolling, but can also occur in other processes. Figure 10.13 shows an example of an alligator crack in a cold-rolled aluminium alloy. The phenomenon is usually associated with relatively small pass reductions, high roll friction and midplane porosity and/or segregation. Just after roll exit, the surface material stays in contact with the rolls and the material in the centre of the workpiece is subjected to very high through-thickness tensions.

If the pass reduction sequence is changed then it is possible to avoid

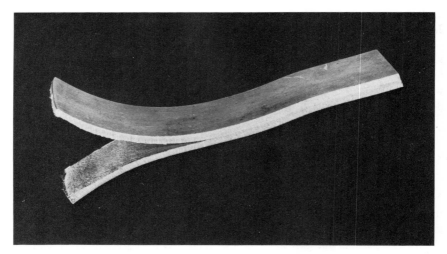

Fig. 10.13 Alligator cracking in an aluminium alloy.

alligatoring. In these cases a series of heavy reductions would normally be adequate.

Because this type of cracking is a function of the material, reduction and friction, it is a complex problem and so far there is no theoretical approach to it.

10.4 Extrusion and drawing

Surface defects such as checkmarks on rod and wire can usually be traced back to die wear or bad or uneven lubrication. Many of the common defects

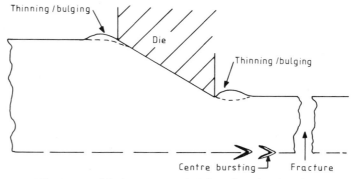

Fig. 10.14 Different possible fractures and defects in wire drawing or containless extrusion of rod.

and fractures that can occur in the manufacture of rod were reported by Jennison (1930).

The chief defects that occur in drawing and free (or containless) extrusion are: (1) fracture of the thinned rod caused by the application of too great a force; (2) bulging of metal at entrance or exit to the dies; (3) thinning or waisting of the metal at the die entrance or exit; (4) centre bursting. These defects are shown diagrammatically in Fig. 10.14.

10.4.1 Application of slip-line theory and upper-bound technique to cracking and central bursting

Much experimental work has been carried out on central bursting in drawing. Because of the possibility of the use of slip-line field analyses, Rogers and Coffin (1971) studied in great detail midplane cracking in plane-strain sheet drawing. From an extensive number of slip-line field calculations they showed in many cases that when the midplane hydrostatic stress becomes significantly tensile this coincides with the onset of centre bursting. Figure 10.15(a) shows a typical two-centred fan field. It was shown by Dodd and Scivier (1975) and Kudo et al. (1972) that some of these slip-line field solutions are statically inadmissible owing to overstressing at either die entry or die exit. Figure 10.15(b) shows the regions of inadmissible slip-line fields and the zone over which bulging and thinning occur. Clearly, whatever the relatively sophisticated arguments about the admissibility of these two-centred fan slip-line fields, to ensure the absence of any centre cracking, hydrostatic tension on the midplane should be avoided by judicious choice of die angle and reduction.

Prompted by a suggestion by W. Johnson (1959 private communication to H. Kudo), Dodd and Kudo (1980) proposed a slip-line field that allows for a form of central splitting in wire drawing. This field is shown in Fig. 10.16. Unfortunately, however, it has been found that this field has limited admissibility.

Avitzur (1977) has applied the upper-bound technique to the onset of centre bursting. Using curved velocity discontinuities, Avitzur compared velocity fields with and without centre bursting. Centre bursting will occur when the energy required for flow with bursting is less than that for flow without bursting. It was found by Avitzur that, as the die angle increases, the velocity discontinuities approach each other as shown in Fig. 10.17. According to Avitzur, for some values of reduction and friction the discontinuities touch, and when this occurs fracture results. It is sufficient to compare the pressures required for fracture initiation with those assuming no fracture. Results are summarized in Fig. 10.18. In each domain in Fig. 10.18

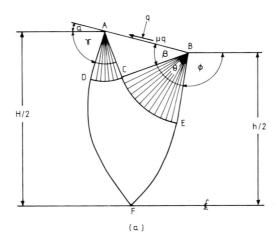

(a)

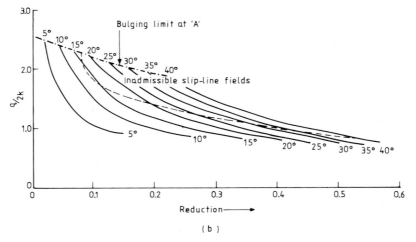

Reduction⟶

(b)

Fig. 10.15 (a) Two-centred fan slip-line field for drawing and extrusion. (b) Regions of inadmissible slip-line fields due to either overstressing at discontinuities A or B. (After B. Dodd and D. A. Scivier, *Int. J. Mech. Sci.* **17**, 663–667 (1975).)

the drawing stress for specific class of velocity field is lower than for any other flow pattern.

For the solution involving two curved velocity discontinuities shown in Fig. 10.17(a), for a given set of process variables there is an optimum die angle α_{opt} for which the force is a minimum. It was found by Avitzur that the

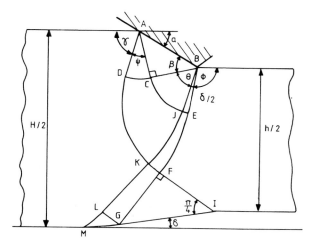

Fig. 10.16 Slip-line field allowing for midplane splitting. (After B. Dodd and H. Kudo, *Int. J. Mech. Sci.* **22**, 67–71 (1980).)

optimum die angle is given approximately by

$$\alpha_{opt} \approx \left[\frac{3}{2}\, m \ln\left(\frac{R_0}{R_f}\right)\right],$$

where m is the friction factor and R_0 and R_f are the initial and final radii of the workpiece.

If the die angle is too large then there is a possibility of dead-zone formation, and the effective die angle is less than the actual angle, as shown in Fig. 10.18. Avitzur presented the following equation for the critical angle that causes dead-zone formation:

$$\alpha_1 = \left[\frac{3}{2}\ln\left(\frac{R_0}{R_f}\right)\right]^{1/2}.$$

If α is less than α_1 then flow is according to Fig. 10.17(a). At very high cone angles a type of machining called shaving occurs. Here the surface layer of the workpiece is shaved off the rest of the rod, which traverses the deformation zone without increasing its velocity. In other words, the core of the material moves through the die without deforming, as shown in Fig. 10.18. The velocity field for centre bursting is shown in Fig. 10.17(c). Here the plastic zone is an annulus, the plastic region not reaching the axis of symmetry. The velocity field is such that as deformation proceeds the size of the central burst increases. Further details of this solution are given by Avitzur (1968).

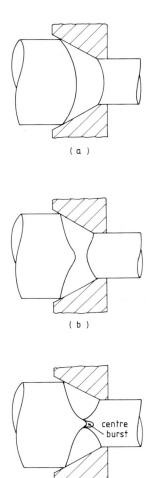

Fig. 10.17 For a range of die angles and reductions in axisymmetric drawing and extrusion the two velocity discontinuities shown in (a) may curve as shown in (b) and overlap, causing a possible centre burst (c). (After B. Avitzur, *Ann. Rev. Mat. Sci.* **7**, 261–300 (1977).)

In all velocity fields that predict some kind of cavity formation, the actual work of cavity generation is ignored except in the case of some of Avitzur's work. Here a work-rate term for pore formation, $\dot{W}_{\text{pore}}$, is included. However, it is not clear whether this modification to the respective extremum principle is valid. The slip-line field suggested by Dodd and Kudo (1980), although

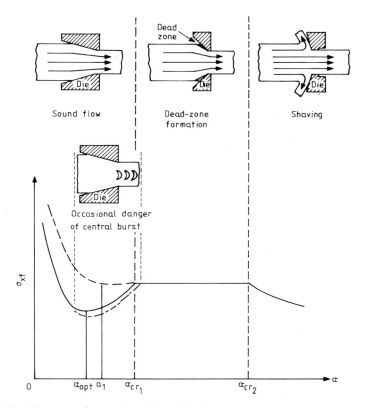

Fig. 10.18 The zone of centre bursting. (After B. Avitzur, *Ann. Rev. Mat. Sci.* **7**, 261–300 (1977).)

predicting centreplane fracture, ignores the energy required to form the new surfaces. No ductile fracture criterion is used in any of these velocity fields, and it may be that a velocity field plus a fracture criterion may be a valid future approach.

The extrusion defect is the formation of a pipe. The final part of extrusion corresponds to a rapid increase in load, and material begins to lift off the punch, forming a pipe along some percentage of the interior of the rod. This pipe is called the extrusion defect, and is a characteristic of the inhomogeneous plastic deformation that occurs in the process.

Kudo has suggested an upper-bound solution that would allow for cavity formation on the ram. This solution for plane-strain extrusion is similar to those for central bursting. An analogous solution may be found for the axisymmetric case, but it is more complicated to derive.

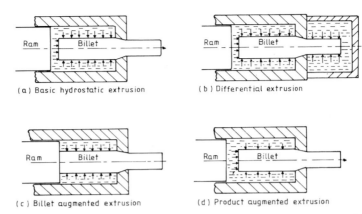

Fig. 10.19 Systems of hydrostatic extrusion and drawing. (After T. Z. Blazynski, *Metal Forming—Tool Profiles and Flow* (1976). By permission of Macmillan, London and Basingstoke.)

10.4.2 Hydrostatic extrusion

We know of the advantageous effect of superimposed hydrostatic pressure on material ductility from Chapter 3. A number of researchers have developed processes such as hydrostatic extrusion and drawing that utilize these advantages. The four systems of hydrostatic extrusion are shown diagrammatically in Fig. 10.19. Basic hydrostatic extrusion uses only the pressure exerted by the oil, while differential extrusion involves extruding from one pressure vessel to another by differential pressure. The other two types of extrusion are augmented extrusion, either by the billet or the product. In the

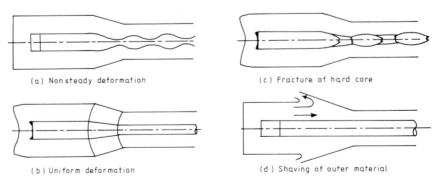

Fig. 10.20 Possible deformation modes in composite rod extrusion, after Osakada *et al., Int. J. Mech. Sci.* **15**, 291–307 (1973).)

latter case the product is drawn through the die, and to this extent it is a combined hydrostatic-extrusion–drawing process.

Hydrostatic extrusion is more appropriate than conventional extrusion for the production of composite rods because the billet is deformed more uniformly. However, the process is complicated because both of the materials have different mechanical properties. As a result of this, there are a number of different possible modes of nonuniform deformation, as shown in Fig. 10.20.

11

Metal Cutting and Related Processes

Unlike the last two chapters, which have been concerned with the forming of metals without material loss, this chapter will deal with the other extreme in the spectrum; that is, the removal of metal from a workpiece to achieve either a designed shape or a required finish. There are several processes of metal removal used in manufacturing, each with different objectives. Section 11.1 outlines some of the processes, along with their main features from the point of view of ductile fracture. Perhaps the most commonly discussed and typical process of metal removal is metal cutting. Johnson and Mellor (1973), Boothroyd (1975), Trent (1977) and Shaw (1984), give comprehensive phenomenological and practical descriptions of the process. In §11.2 we shall describe the major aspects of metal cutting and their relationships to ductile fracture.

Idealized two-dimensional metal cutting—so-called orthogonal cutting—provides a realistic and helpful model, since "a completely general numerical approach is beyond even modern computing machines" (Hodge 1970). Although it can be argued that present-day finite-element codes can handle three-dimensional machining processes to some extent; two important problems still require solution: (i) a criterion is needed to determine for the program when, where and how a chip is formed; (ii) remeshing techniques of sufficient generality are required to allow a study of chip formation to be carried out in a practical time. Hence §11.3 is devoted to the study of models of orthogonal cutting, with special emphasis on the facets of ductile fracture involved and the application to this field of theories from Chapters 4–7. Finally some remarks are made about blanking as a comparison with the cutting analysis.

11.1   Main features of processes relevant to metal removal

In manufacturing there are a variety of cutting operations (see Fig. 11.1). Turning uses a single-point tool on a lathe to remove material and produce a new surface of revolution, while boring produces a new internal surface of revolution. Milling uses a multipoint tool to produce plane or curved surfaces. Drilling is another widely used process for removing unwanted material and producing a hole in the workpiece. Owing to the shape of the hole, the drilling tool is complex, although the cutting operation, usually with two cutting edges, is similar to the previous operations. Similar processes, but for special usage, are sawing, reaming, tapping and threading (Fig. 11.1). As with drilling, each process is used to produce a required shape or finish; hence the tool is specially designed and usually complex in shape. However, the essential features remain the same, namely removal with a cutting edge of part of the metal near the surface to be machined. The operations are governed by the following factors: cutting speed, feed rate, the geometry of the cutting edge (such as rake angle and inclination angle), the surface to be machined, the mechanical behaviour of the tool, the work material, as well as friction and lubrication between tool and workpiece. Grinding and polishing should be mentioned here because they are quite similar to the aforementioned processes, except that there is an absence of any apparent sharp cutting edges. Finally, a different group of metal-removal processes are those in which the workpiece is separated into two, in processes like blanking, punching, cropping and guillotining (Fig. 11.2). Compared with cutting, these processes consist of comparatively simple tools and workpiece geometries; therefore the main feature of separation by these processes is dominated by shear, either sliding (as in blanking) or tearing (as in guillotining), together with cracking. Certainly, a careful study should take account of the accompanying phenomena, for example (as shown in Fig. 11.2b) draw-in, tilting and burnished-band formation.

All of the metal-removal operations possess one feature in common, namely the creation of new surfaces of the workpiece, rather than changing the shape of the original surface as is done in either sheet or bulk forming. Furthermore, both sheet and bulk forming must be performed within a certain forming limit, to which ductile fracture is related. On the other hand, metal cutting should be carried out beyond the threshold defined by ductile fracture, even though there are both localization of finite plastic deformation and fracture occurring. Therefore, from the point of view of ductile fracture, both metal-forming and cutting processes share the same theoretical background.

Because of the complexities involved in metal cutting, there have only been some basic ductile fracture theories that have been used to deal with

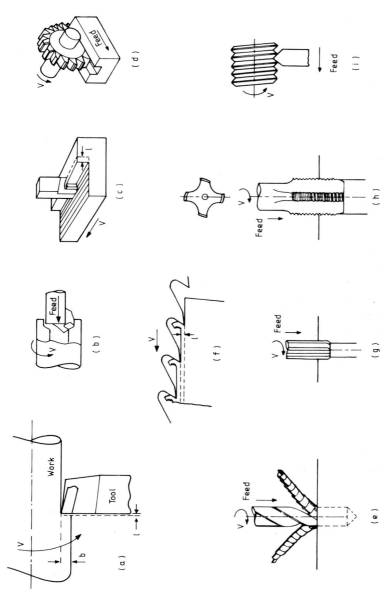

Fig. 11.1 Miscellaneous cutting operations: (a) single-point turning, (b) boring, (c) planing, (d) milling, (e) drilling, (f) sawing, (g) reaming, (h) tapping, (i) threading. (From M. C. Shaw, *Metal Cutting Principles*. Clarendon Press, Oxford, 1984.)

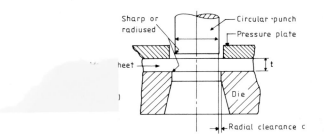

Sharp or radiused

Circular punch

Pressure plate

Sheet

Die

t

Radial clearance c

(b)

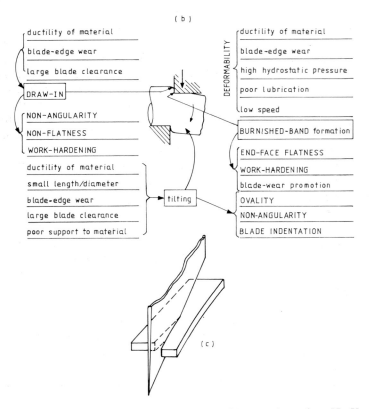

ductility of material

blade-edge wear

large blade clearance

DRAW-IN

NON-ANGULARITY

NON-FLATNESS

WORK-HARDENING

ductility of material

small length/diameter

blade-edge wear

large blade clearance

poor support to material

DEFORMABILITY

ductility of material

blade-edge wear

high hydrostatic pressure

poor lubrication

low speed

BURNISHED-BAND formation

END-FACE FLATNESS

WORK-HARDENING

blade-wear promotion

OVALITY

NON-ANGULARITY

BLADE INDENTATION

tilting

(c)

Fig. 11.2 Shearing operations, (a) blanking, (b) cropping after H. Kudo, (c) guillotining.

simplified cutting problems. Hence this chapter is mainly aimed at properly displaying the characteristics of ductile fracture in these processes. Nevertheless, because these are high-volume processes, even minor improvements in the basic understanding could be of crucial significance and benefit.

11.2 Essential phenomena, concepts and their interactions in cutting

11.2.1 Orthogonal cutting and basic parameters

To obtain a physical feeling for the role of ductile fracture in metal cutting, only two-dimensional orthogonal cutting is considered here in detail, theoretically and experimentally. If chip formation in cutting is two-dimensional then the cutting process is referred to as orthogonal cutting. A single-point turning is usually a three-dimensional operation, but it can be treated as orthogonal cutting, provided that the cutting edge is perpendicular to the cutting velocity, and the width b of the cutting edge is large enough, usually five or more times the feed, i.e. the depth of cut or undeformed chip thickness (Fig. 11.1a). Clearly, orthogonal cutting is a plane-strain process (Fig. 11.3). The first impression of cutting is that the chip moves along the rake, the chip is thicker than the feed and there is a high temperature and large finite deformation and fracture of the chips.

   The tracings of a deformed grid from a quick-stop orthogonal cutting test are shown in Fig. 11.4. There is a distinct transition region in the vicinity of line AB, revealing severe distortion; hence this is named the primary shear zone. Another severely distorted region is close to line AC, i.e. the interface of the chip and the rake of the tool. This is the so-called secondary shear zone. The rest of the workpiece seems to remain undeformed. All these deformation features are sketched in Fig. 11.3(a). Figure 11.3(b) shows the forces exerted on the chip. Representative data for metal cutting show that the surface energy due to new surface formation and the momentum change while the metal crosses the shear plane are negligible compared with the shear energy $U_s = F_s V_s$ in the primary shear zone and the friction energy $U_f = F_c V_c$. In addition, the ratio U_s/U is in the range 0.5–0.3, according to Shaw (1984), where $U = F_p V$ is the total energy consumption per unit time.

   For steady orthogonal cutting, it is clear from Fig. 11.3 that there are several parameters that dominate the cutting process (typical values are given in parentheses):

geometrical: rake angle ($-10°$ to $+10°$);

t, feed or cutting depth or underformed chip thickness (0.01–1 mm);

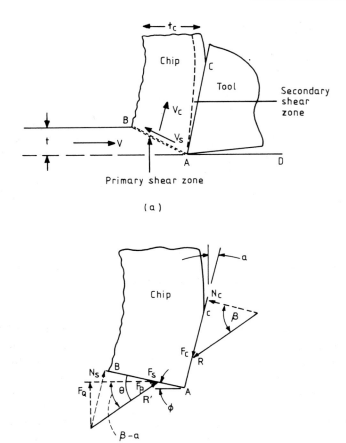

Fig. 11.3 Orthogonal cutting: (a) the orthogonal cutting operation, (b) free-body diagram of the chip.

kinematic: v, cutting velocity (30–300 m min^{-1})

physical: friction on the interface of the chip and rake—to some extent, the friction coefficient μ or the friction angle $\phi = \tan^{-1} \mu$ could be representative of friction; work-material behaviour—perhaps this is one of the most complex factors in cutting, in addition to ductile fracture.

The behaviour of the work material should encompass finite and localized plastic flow, heat and temperature rise, inhomogeneities and fracture. Hence

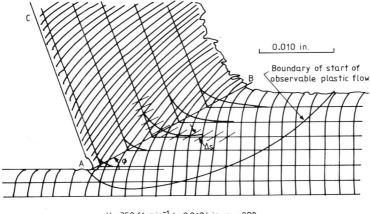

V = 350 ft min^{-1}, t = 0.0104 in., α = 20°

Fig. 11.4 Tracing of deformed grid from quick-stop test. (After M. Stevenson and P. L. B. Oxley, *Proc. Inst. Mech. Engrs* **185**, 55–71 (1970). Reprinted by permission of the Council of The Institution of Mechanical Engineers.)

important parameters may include shear-flow stress k, strain-hardening exponent n, characteristic temperature T_h and strain rate $\dot{\gamma}_h$, heat diffusion and volume specific heat, a microstructural lengthscale a and some fracture criterion. Other factors of importance in the cutting process are the shear angle ϕ, the chip thickness t_c and velocity V_c, stress and strain state (especially the horizontal force component F_p), the energy consumption, and the chip shape. The following subsections will describe some qualitative interrelationships of the dominant parameters that affect cutting, but the emphasis is on material behaviour rather than the mechanics of cutting.

Before addressing these problems, it is necessary to observe that the actual picture of cutting is more complicated than has been described. This is not only due to the steady and orthogonal simplification, but also to the complexity caused by tool wear, the extended plastic zone (Fig. 11.4), discontinuous chips (leading to unsteady flow), elasticity of the machine, and so forth.

11.2.2 Primary and secondary shear zones

Let us examine Fig. 11.3(a) again. The newly created surface AD is parallel to the cutting velocity. Another new surface attached to the chip is forced to travel upward and coincides partially with the rake surface AC. Furthermore, the chip, i.e. the material confined by CAB, is under loadings R and R' from the tool and workpiece as shown in Fig. 11.3(b). This pattern could be a mixture of fracture modes, i.e. the opening mode (tension) and the sliding mode (shear). However, the chip is constrained to move along the surface of the rake, rather than in the direction of AD. Hence intense shear, unlike that in the sliding mode, is inclined to the cutting velocity V with a shear angle ϕ. The conclusion is that primary shear is a prerequisite for the production of the new surface AD in cutting. The question of fracture mode will be left until §11.2.4 dealing with chip formation. As for the secondary shear zone, it appears to be mainly caused by friction; hence it normally appears in high-speed cutting and is accompanied by high temperatures and wear.

Historically it is interesting to consider what is called Ekstein's paradox of cutting. That is, as Drucker (1949) quoted, "the stresses are extremely high, of the order of the ultimate static strength, in a large region of the chip and workpiece, and yet the zone of shearing is extremely narrow". If one recalls the foregoing sections on localization of plastic deformation, especially on the various mechanisms of shear banding (Chapter 6 and §§7.1.2 and 7.5.6), the formation of the primary shear band is a simple matter to understand. Beyond a certain threshold, uniform deformation must give way to a narrow shear band, and the line AB is located at the transition from velocity V of the workpiece to the chip velocity V_c.

The measurement of the shear zone by Stevenson and Oxley (1970) with an explosive quick-stop device shows the details of the primary shear zone clearly and validates the above picture. The distribution of the maximum shear-strain rate is roughly symmetrical to the shear line AB, and along AB the strain rate remains nearly constant, but with a slight increase in the vicinity of point A.

Now it is necessary to define relevant variables in the primary shear zone. From Fig. 11.5(a) the deformation near AB can be treated either as a variation from ABIJ to ABDC or as simple shear from ABGE to ABDC. If the displacement of ABIJ is neglected then the shear strain is

$$\gamma = \frac{CF + FE}{AF} = \tan(\phi - \alpha) + \cot\phi$$

$$= \frac{\cos\alpha}{\sin\phi\cos(\phi - \alpha)}. \tag{11.1}$$

The variation of γ with ϕ and α shows a tendency for the shear strain to increase with decreasing rake angle and shear angle. Typical values for the shear angle are between $10°$ and $40°$, and this gives a shear strain of between approximately 1 and 10. From Fig. 11.5(b), the strain rate is given by

$$\dot\gamma = \frac{V_s}{AF} = \frac{\cos\alpha}{\cos(\phi - \alpha)}\frac{V}{\delta}, \tag{11.2}$$

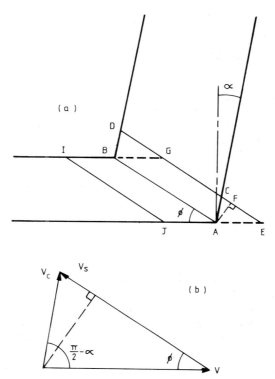

Fig. 11.5   Determination of shear strain: (a) shear strain in orthogonal cutting, (b) velocity diagram. (After G. Boothroyd, *Proc. Inst. Mech. Engrs.* **177**, 789–802 (1963). Reprinted by permission of the Council of the Institution of Mechanical Engineers. See also G. Boothroyd, *Fundamentals of Metal Machining and Machine Tools.* McGraw-Hill, New York, 1981.)

where $\delta = CF$ and denotes the width of the primary shear zone. If we take Drucker's estimate, $\delta \sim t/20$ and if we suppose that $V_s \approx 30$ m min^{-1} and $t \approx 0.1$ mm then $\dot{\gamma} \sim 10^5$ s^{-1} and the time for the shear process is 10 μs. This rate and timescale are fairly unusual industrially. Another important feature of the shear zone is its temperature, which is left until the next subsection for description.

There are two variables characterizing the primary shear zone; these are the shear angle ϕ and the width δ. The shear angle ϕ clearly determines V_s and V_c, which affect the whole deformation pattern. The cutting ratio r is given by

$$r = \frac{\text{feed}}{\text{chip thickness}} = \frac{t}{t_c} = \frac{AB \sin \phi}{AB \cos (\phi - \alpha)} = \frac{\sin \phi}{\cos (\phi - \alpha)}, \qquad (11.3)$$

or

$$\tan \phi = \frac{r \cos \alpha}{1 - r \sin \alpha}.$$

Therefore the shear angle ϕ is a "bridge" variable, which connects the predetermined parameters with the result of the process of cutting. This again shows the dominant role of primary shear in metal cutting.

Here we shall briefly examine the factors that govern the shear angle ϕ. Based on the list at the beginning of the section, ϕ should be

$$\phi = f(\alpha, t, v, \text{friction, work material}).$$

To simplify this, suppose that the friction angle β, the flow stress k and the strain-hardening exponent n are enough to represent friction and the work material respectively; then one can deduce from dimensional analysis that

$$\phi = f(\alpha, t, v, \beta, k, n)$$

$$= F(\alpha, \beta, n). \qquad (11.4)$$

This implies that the shear angle should be independent of feed and cutting velocity. This result is certainly interesting, but clearly has limited usage.

If a perfectly plastic work material is assumed then the parameter n will vanish from (11.4); then

$$\phi = F(\alpha, \beta). \qquad (11.5)$$

Therefore, in order to obtain a better relation, for instance making ϕ dependent on feed and cutting velocity, one has to seek other parameters, which should contain the dimensions of time or length, from the work material and friction. According to the significance of the primary shear zone with its high strain rates and strong localization, the study of plastic localization and relevant ductile fracture are of crucial importance in metal cutting. At the moment, the fact that greater rake angle leads to a greater shear angle is an acceptable rule from a geometrical point of view.

11.2.3 Temperature rise

Figure 11.6 shows a typical temperature distribution in orthogonal machining, derived from infrared photography. There are two regions in which the temperature and its gradient are high, and they coincide with the primary and secondary shear zones. The shear-plane temperature will definitely affect the mechanical behaviour, such as the flow stress of the work material within the primary shear zone.

It is safe to assume that almost all of the plastic energy is converted into

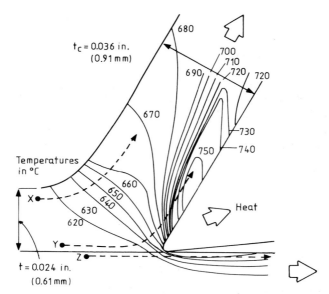

Fig. 11.6 Temperature distribution in workpiece and chip during orthogonal cutting (obtained from an infrared photograph) for free-cutting mild steel where the cutting speed is 75 ft min^{-1} (0.38 m s^{-1}), the width of cut is 0.25 in. (6.35 mm), the working normal rake is 30°, and the workpiece temperature is 611°C. (After G. Boothroyd, *Proc. Inst. Mech. Engrs* **177**, 789–802 (1963). Reprinted by permission of the Council of the Institution of Mechanical Engineers.)

heat, but the heat in metal cutting is conducted away by two means. One portion is transferred to the tool and the workpiece and the other portion is convected away by the chip. Consequently, the average temperature rise in the shear plane, $\bar{T}_s$, can be written as

$$\bar{T}_s = \frac{\int \tau \, d\gamma}{\rho c_v} (1 - \eta), \qquad (11.6)$$

where η represents the fraction of heat transferred to the chip, tool and workpiece while shearing. Accordingly η should be a function of a dimensionless variable, which should combine the thermal and kinematic processes in cutting. Since the thermal conductivity κ must be involved in the variable, η would have the form

$$\eta = f\left(\frac{l_*^2}{\kappa t_*}\right), \qquad (11.7)$$

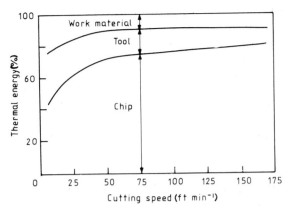

Fig. 11.7 Variation of energy distribution with drilling speed. (After A. O. Schmidt and J. R. Roubik, *Trans. ASME* **71**, 245–248 (1949).)

where t_* and l_* are the characteristic time- and lengthscales. In the case of cutting, l_* should be the width of the shear zone, and because of conservation of mass $Vt = V_s \delta$. This leads to

$$\frac{l_*^2}{\kappa t_*} \simeq \frac{Vt}{\kappa} \frac{\delta}{t_* V_s} = \frac{Vt}{\kappa} \frac{1}{t_* \dot{\gamma}_s} = \frac{Vt}{\kappa} \frac{1}{\gamma_s}. \tag{11.8}$$

$R^* = Vt/\kappa$ is called the thermal number; it combines thermal conductivity κ with convection. It is quite clear from (11.1) that using cot ϕ to describe the shear strain is not a bad approximation in the case of low shear angles. Both Boothroyd's (1963) data and the two other simplified models, due to Loewen and Shaw (1954) and Weiner (1955), give a decreasing function of η with increasing $(Vt/\kappa) \tan \phi$.

Interestingly, a large amount of work-material flow (Vt) can offset the thermal conduction, decrease η and maximize the shear-plane temperature. The portion of energy would eventually be taken by the chip. Figure 11.7 shows typical variations of energy with cutting speed. It is possible to visualize the significance of the temperature rise in the shear zones, and therefore the formation of the chip. Incidentally, because of heat transfer, both the cutting velocity and the depth of cut must be included in the dimensionless governing parameter.

11.2.4 Chip formation

Chip formation is the key to metal cutting, and is directly related to the primary shear zone and fracture.

There are two types of chip: a continuous ribbon and individual segments. Continuous ribbons are always ductile, but in some cases one can see inhomogeneous shear and even partially ductile shear fracture (see Fig. 11.8). All of these occur at the primary shear zone; therefore they are a type of localized shear. As has already been discussed in Chapters 6 and 7, there are two forms of localization—these are heat-assisted shear localization and the isothermal plastic localization. The former occurs with a material that has a low thermal conductivity and specific heat; therefore the temperature rises readily and there is a tendency to softening at elevated temperatures.

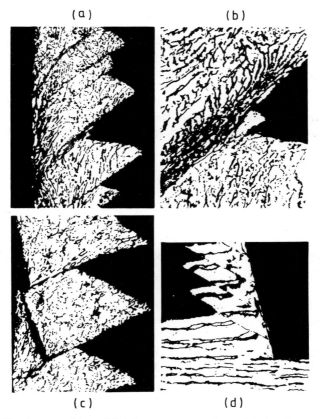

Fig. 11.8 Continuous chips with inhomogeneous shear: (a) titanium cut at a high speed (53 m/min^{-1}), adiabatic shear; (b) enlargement of (a); (c) titanium cut at low speed (25 mm min^{-1}), periodic fracture, gross sliding and rewelding; (d) 60–40 cold-rolled brass (60% reduction in area) cut with HSS tool having minus 15° rake angle. Cutting speed 0.075 m min^{-1}; undeformed chip thickness 0.16 mm. (Sketches based on a figure in M. C. Shaw, *Metal Cutting Principles*. Clarendon Press, Oxford, 1984.)

Titanium alloys are typical of this type of material (Figs. 11.8a, b). In the second case, thermal softening seems not to be significant in shearing; hence the width of the zone of high shear strain is comparatively narrow and perhaps accompanied by shear fracture (Figs. 11.8c, d). Individual segments are produced by fracture, but the gross fracture surface of each segment generally coincides with the primary shear zone, and some segments can contain some plastic deformation. Experiments show that the transformation from a discontinuous chip to a continuous chip occurs when the rake angle is increased. Hence, generally speaking, a tendency to form a discontinuity increases with decreasing rake angle (therefore usually decreasing the shear angle) and cutting velocity, and increasing the depth of cut and friction. Incidentally, the so-called built-up edge (BUE) on the tool rake usually appears at comparatively low cutting speeds, and this will correspond to an increase in the rake angle, hence making the chip continuous. From the point of view of ductile fracture, it is necessary to find which material factors govern the transition between the two types of chip. Figure 11.9 shows the variation of shear stress with normal stress on the fracture surface. The × in Fig. 11.9 identify fracture of segmental chips and form a line inclined to the axis of normal stress; beneath this line (high rake angles 30° and 45°) chips remain

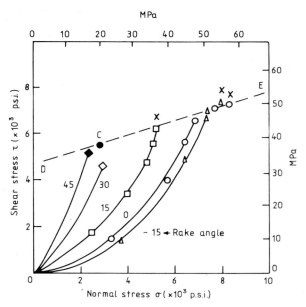

Fig. 11.9 Variation of shear stress with normal stress on fracture surface when cutting. (After Cook *et al.*, *Trans. ASME* **76**, 153–162 (1954).)

continuous. This reveals that the fracture is governed by the normal stress. Some relevant results given in §§2.3.3 and 3.1.1 and Fig. 3.1 have shed light on the problem, showing that the shear strain to fracture or the maximum shear stress (Fig. 3.1) increase with increasing hydrostatic pressure. A direct study is shown in Fig. 11.10, in which the effect of normal compressive stress on the maximum shear stress and strain becomes obvious. Hence the underlying factor controlling the transition from a continuous chip to a discontinuous chip is the shear strain at fracture, which depends on the normal stress on the shear plane (see also Sharma *et al.*, 1971).

As shown in Figs. 11.3(b) and 11.9, the shear plane, on the whole, is under the combined stress state of shear and compression. If this were true everywhere in the shear plane then the workpiece would hardly be separated to form new surfaces and a chip, even though the fraction of the energy for the formation of the new surface is negligible (§11.2). So there should be a certain tensile fracture somewhere in the process. Historically, there were two models

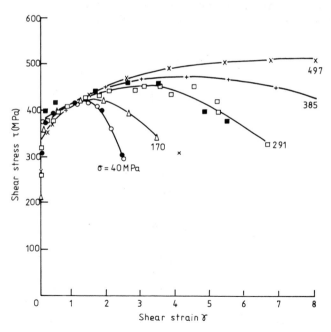

Fig. 11.10 Shear-stress–shear-strain results for resulphurized low-carbon steel, where σ is the normal stress on the shear plane. (After T. J. Walker and M. C. Shaw, in *Proc. 10th Int. Machine Tool Design Research Conf.* (ed. S. A. Tobias and F. Koenigsberger), pp. 241–252. Pergamon Press, Oxford, 1969.)

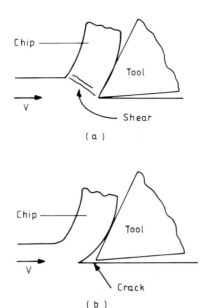

Fig. 11.11   Models of cutting: (a) model with shear, (b) model with cracking.

of cutting, as shown in Fig. 11.11. One stressed the significance of shear in cutting (Fig. 11.11a) and is now generally accepted. But the original model mentioned nothing about the formation of the new surface; hence an alternative model found quite wide support for many years, which suggested that a crack occurred ahead of the tool tip (Fig. 11.11b). Since the second model discounts shear, a vital factor in both the energy consumption and the deformation pattern, it is considered to be a misconception. So far the question of new surface formation has remained open. It seems that the creation of the new surface may occur in a very local region near the tool tip. In that region the finite curvature of the tool tip has to be taken into account, no matter how sharp the tool is. Iwata *et al.* (1984) carried out a study of the problem with the finite-element method and microscale cutting within a scanning electron microscope. The computational results are shown in Fig. 11.12. In Fig. 11.12(a) a distinct narrow shear zone coincides with the primary shear, which is the basis of the first model (Fig. 11.11a), whereas in Fig. 11.12(b) the distribution of hydrostatic stress is shown, and a tensile stress does exist and concentrate in the vicinity ahead of the tool tip. A ductile fracture criterion similar to (8.6) in which the hydrostatic tension and voids play a significant role in fracture has been adopted in computation, and some

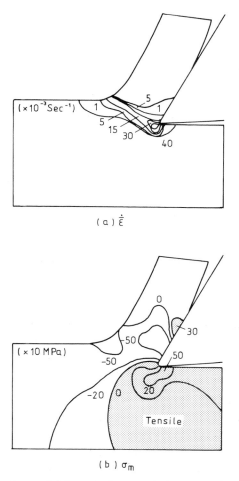

Fig. 11.12 Distributions of (a) equivalent strain rate and (b) hydrostatic stress during orthogonal cutting calculated by FEM ($\alpha = 30°, t = 50\ \mu m, r = 5\ \mu m, \mu = 0.5, v = 1\ \mu m\ s^{-1}$), where r is the radius of the tool edge. (After K. Iwata *et al.*, *Trans. ASME, H: J. Engng Mat. Sci. Technol.* **106**, 132–138 (1984).)

cracks do occur ahead of the tool tip. The cracks observed in SEM photographs correspond well with computational predictions. Therefore cracking seems to be very likely in the very vicinity of the tool tip owing to tension, perhaps in accord with bending or local stretching by neighbouring elements of the chip in the work-material element around the tool tip. In fact, the cracks, if they occur, cannot travel far ahead of the tool tip owing to the main field of shear and compression in the transition area from the workpiece to the chip. The tensile cracks, no matter how small, separate the chip from the workpiece and may act as a trigger to make deformation give way to shear localization. Therefore the primary shear zone, which is the weakest part in the stiffness of the cutting system, offsets most of the stored energy and then dominates the overall pattern of the cutting process.

11.2.5 Specific energy and size effect

Like the size effect in the general discussion of ductile fracture in §1.9, there is a size effect in metal cutting, but it has a special form. As shown in Fig. 11.13(a), the specific cutting pressure, i.e. the force parallel to the cutting velocity in unit area of cut, F_p/tb, or equivalently the specific cutting energy $u = F_p V/(Vbt) = F_p/bt$ is by no means geometrically similar for the same material. To some extent, the size effect can be approximately expressed by

$$u \sim t^{-0.2}. \tag{11.9}$$

This shows that the thinner the depth of cut, the higher the specific cutting energy. According to the discussion in §11.2.2, this should be attributed to some intrinsic length- or timescale involved in the work-material behaviour. But the details, due to different investigators, remain uncertain. Drucker (1949) was perhaps the first to attempt to deal with this significant problem, by considering inhomogeneity and discontinuous plastic deformation. Recently Shaw (1984) has given a quantitative analysis within the same approach.

Supposing that the shear angle is the same for all depths of cut, then the shear strain is constant as well for fixed rake angles. One can deduce that the larger the depth of cut, the more numerous are the planes of weakness, which are randomly distributed in the work material, and the intrinsic lengthscale becomes insignificant. As for the inverse relation (11.9), Drucker suggested that the more planes of weakness that are involved, the less the average shear stress on the shear planes, until the number of planes of weakness become saturated. Oxley and his coworkers (Larson-Basse and Oxley, 1973; Kopalinsky and Oxley, 1984) attributed the size effect to the increase of strain rate in the primary shear zone with a small depth of cut by recalling that elevated strain rates enhance the flow stress (§3.4.1). This is, of course, very likely for cutting, since strain rates as high as 10^5 s^{-1} may be involved. At these rates the flow stress shows an abrupt increase. However, the accompanying high temperature in the primary shear zone may offset the rate effect.

The experimental results of Kopalinsky and Oxley (1984) are shown in Figs. 11.13(a, b). It seems that at large depths of cut (>0.05 mm) the size effect may be due to both an increase in the flow stress and a decrease of the shear angle, but at small depths of cut (<0.05 mm) the dramatic increase in the flow stress (most probably caused by the strain-rate effect) appears to be the major factor governing the size effect. The curves in Fig. 11.13 represent the theoretical predictions based on a machining theory of a rate- and temperature-dependent work material, which will be outlined in the next section.

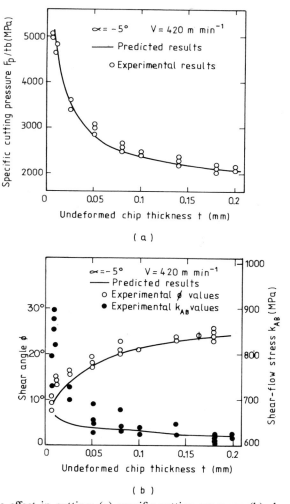

Fig. 11.13 Size effect in cutting: (a) specific cutting pressure; (b) shear angle and shear-flow stress. (After E. M. Kopalinsky and P. L. B. Oxley, in *Mechanical Properties at High Rates of Strain* (ed. J. Harding), pp. 389–396. Conf. Ser. No. 70, Inst. Phys., London, 1984.)

11.3 Mechanical models of orthogonal cutting based on localized shear deformation

This section will deal with several models to determine the interrelationship of various parameters in cutting. The emphasis is still on material behaviour; therefore the primary shear zone and the related behaviour of plastic deformation and fracture are naturally the focus of the section.

11.3.1 Piispanen's card model

Piispanen (1937) first put forward the card model, which exaggerates the key facet of metal cutting, i.e. *concentrated simple shear*. Once the metal reaches the shear plane AB (Fig. 11.14), it is sheared without any subsequent further distortion. Hence the chip appears to be a deck of cards inclined to the cutting velocity by shear angle ϕ. This model, although oversimplified, does provide a very valuable description of cutting.

11.3.2 Maximum-shear-stress and minimum-power criteria

Ernst and Merchant (1941) presented what appears to be the first quantitative theory based on the card model and the criterion of maximum shear stress. Supposing that the shear stress τ on the shear plane is uniform (see Stevenson and Oxley, 1970), one can obtain (referring to Fig. 11.3b)

$$\tau = \frac{F_s}{A_s} = \frac{R \cos (\phi + \beta - \alpha)}{A \operatorname{cosec} \phi}, \tag{11.10}$$

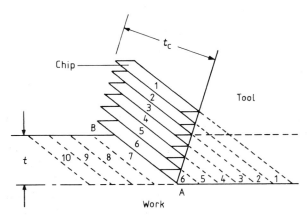

Fig. 11.14 Piipanen's idealized model of cutting.

where A_s and A are the areas of the shear plane and the section of the depth of cut respectively. $d\tau/d\phi = 0$ leads to

$$\phi = \tfrac{1}{4}\pi - \tfrac{1}{2}\beta + \tfrac{1}{2}\alpha. \tag{11.11}$$

Here one has to note that only under the condition of R/A and β being independent of ϕ, i.e.

$$\frac{d(R/A)}{d\phi} = \frac{d\beta}{d\phi} = 0,$$

$$\text{can} \quad \frac{d\tau}{d\phi} = 0$$

be consistent with maximum-shear-stress criterion. From the point of view of the energy consumed, Merchant (1945) gave

$$W = F_p V = R \cos(\beta - \alpha)V = \frac{\tau A \cos(\beta - \alpha)}{\sin\phi \cos(\phi + \beta - \alpha)}.$$

Similarly $d(W/A)/d\phi = 0$ leads to the same expression as (11.11), provided that τ and β are both independent of ϕ for fixed α. Although (11.11) is concise, it has not been borne out by experiments, as shown by Pugh (1958). This is due to the fact that the deduction involves too many uncertain assumptions (Shaw, 1984). However, the underlying fact is that the deduction has not placed enough emphasis on the intense localized shear deformation in the primary shear zone and on the work-material behaviour there (see §11.2.2).

11.3.3 Slip-line field solutions

The observed data on shear angle ϕ can be fitted to a straight line

$$\phi = \text{constant} - \beta + \alpha. \tag{11.12}$$

This can be partially achieved by slip-line field theory, as shown by Lee and Shaffer (1951).

As mentioned in §7.1.4, slip-lines are characteristics across which the tangential velocity is discontinuous. Unlike in the last section, slip-line field theory exaggerates the concentrated shear deformation in the primary shear zone. Therefore it may lead to a better prediction.

As stated in §7.1.4, slip-lines are two families of discontinuities orthogonal to each other for a perfectly plastic homogeneous solid deformed in plane strain. The primary shear zone should coincide with one of these characteristics; in addition, there is no further deformation outside the zone. Hence the slip-line field should be as shown in Fig. 11.15(a), and region ABC is of uniform stress and can be expressed in a unique Mohr circle (Fig. 11.15b).

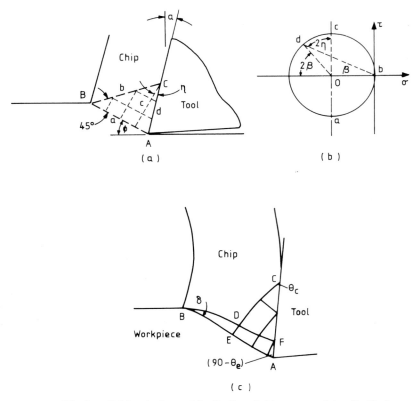

Fig. 11.15 Slip-line field solutions: (a) slip-line field proposed by E. H. Lee and B. W. Shaffer (1951); (b) Mohr's circle diagram for field (a); (c) slip-line field with nonuniform friction between chip and tool, suggested by H. Kudo (1965).

Since AB is a plane of maximum shear stress, the line BC, which is inclined at 45° to AB, should be stress-free, corresponding to point b in Fig. 11.15(b). The stress on line AC can be derived in two ways. From Fig. 11.15(a) $\eta = \phi - \alpha$, and from friction loading (Fig. 11.15b), $\eta = \frac{1}{4}\pi - \beta$. Therefore it follows that

$$\phi = \tfrac{1}{4}\pi - \beta + \alpha, \tag{11.13}$$

which is similar to the line derived experimentally (11.12). However, the intersection of the line is not always the same. So it may be that slip-line field theory is oversimplified for this case. In fact the basis of slip-line field theory is rigid perfect plasticity, wherein only one material parameter, the flow stress k, has been taken into account. As shown in the expression of (11.5), it is impossible to anticipate more from the above slip-line field solution. Hill (1954) suggested that owing to the lack of geometrical constraint in metal cutting, many other solutions are permissible. Additionally, there are a number of variables involved in metal cutting. But quite clearly there are two possible approaches to a more satisfactory solution; namely, modifying the description of friction and taking into account the ductile behaviour of the work material within the primary zone.

K udo (1965) proposed a number of slip-line fields for cases of nonuniform friction-stress distributions between the chip and tool rake face. The most complicated one is shown in Fig. 11.15(c). There are two overlapping regions: the primary shear zone AEBDF and AFCDE adjacent to the rake face. There are three parameters δ, θ_e and θ_c concerning the friction-stress distribution on the rake face. If $\delta = 0$ then $\theta_e = \theta_c$ and the field will be reduced to Lee and Shaffer's solution. Kudo's solutions predict the chip shape reasonably well, as shown by Childs *et al.* (1972).

Oxley took into account the effect of strain hardening, which might be of great importance in cutting. As stated before, however, consideration of the strain-hardening exponent is not enough for the description of work-material behaviour within the primary shear zone, which, most probably, is rate- and temperature-dependent. In addition, introducing the index n still cannot explain the size effect (§7.2.5), as shown in (11.4). Hence in recent years Oxley and coworkers have developed a machining theory assuming the work material to be temperature- and rate-dependent.

11.3.4 Temperature- and rate-dependent theory

In the last ten years Oxley and co-workers have discussed the effects of temperature and strain rate on cutting (Stevenson and Oxley, 1970; Hastings *et al.*, 1980). For carbon steel the strain and temperature in the primary shear zone will be of the order of 2 and several hundred degrees respectively, while in the secondary shear zone the strain will be higher and the temperature will be about $1000\,°C$; the strain rate in both regions is approximately 10^4–$10^6\,s^{-1}$ (Fig. 11.6). Hence the main body of the theory describes the material behaviour under these conditions. After curve fitting, the stress–strain relation for uniaxial stress is expressed as

$$\sigma = \sigma_* \varepsilon^n, \tag{11.14}$$

and the temperature and rate effects are combined into a modified temperature

$$T_v = T\{1 - A \log (\dot{\varepsilon}_p/\dot{\varepsilon}_1)\}, \tag{11.15}$$

where A is a constant.

The variations of σ_* and n with T_v are determined by compression tests on carbon steel under loading similar to cutting. The calculation of the temperature rise is in accord with the fitting of Boothroyd's data (§11.2.3). In the computation there are trial-and-error calculations of temperatures in the primary and secondary shear zones to satisfy the temperature- and rate-dependent constitutive equation, temperature rise and the variation of material parameters with temperature. In addition, some empirical estimates are used in the calculation. The theory has been used to predict the size effect, as the results shown in Fig. 11.13 are in good agreement with experiments. The theory, although semi-empirical, is a good start in dealing with intensely localized deformation and the relevant mechanical behaviour.

11.3.5 Heat-assisted shear localization and segmented chips

In §11.2.4 it was stated that with decreasing cutting velocity the tendency to form discontinuous chips increases. However, in recent years interest in segmented-chip formation has gained impetus as high-speed machining has been used more. This type of semented chip is formed when titanium alloys are cut at nearly all speeds and when steels are cut at quite high speeds. Recently several groups (Semiatin and Jonas, 1984; von Turkovich, 1982) treated this problem as a thermoplastic shear instability.

In fact, shear localization has already appeared in continuous cutting in the form of the primary shear zone. The transition from continuous to segmented chips should correspond to the instability of the steady continuous cutting pattern. Clearly, some attention should be paid while making use of the material-instability approach in this transition.

Semiatin and Jonas (1984) supposed that deformation is uniform over the cross-section of the chip and that the problem could be treated as a one-dimensional instability of the material, $d\tau = 0$ (see §6.4); then

$$0 = \left(\frac{\partial \tau}{\partial \gamma}\right)_{\dot{\gamma}, T} + \left(\frac{\partial \tau}{\partial \dot{\gamma}}\right)_{\gamma, T} \frac{d\dot{\gamma}}{d\gamma} + \left(\frac{\partial \tau}{\partial T}\right)_{\gamma, \dot{\gamma}} \frac{dT}{d\gamma}, \tag{11.16}$$

where $\tau = \tau(\gamma, \dot{\gamma}, T)$, or

$$\frac{A}{\sqrt{3}} = \frac{1}{\dot{\gamma}}\frac{d\dot{\gamma}}{d\gamma} = -\left\{\left(\frac{\partial \ln \tau}{\partial T}\right)_{\gamma, \dot{\gamma}}\left(\frac{1}{\tau}\frac{dT}{d\gamma}\right) + \left(\frac{\partial \ln \tau}{\partial \gamma}\right)_{\dot{\gamma}, T}\right\}/m, \tag{11.17}$$

where $m = (\partial \ln \tau/\partial \ln \dot{\gamma})_{\gamma, T}$. The term $\tau^{-1} dT/d\gamma$ should be determined by heat transfer in cutting; see (11.6). Now the left-hand side of (11.17) is a dimensionless number, while the right-hand side consists of material and cutting parameters. According to the data for AISI 4340 steel, it is found that $A \approx 5$ signifies the formation of serrated chips due to the plastic work converted into heat. Clearly, this is just a preliminary study.

von Turkovich (1982) noticed the difference between the existing shear-localization strains in the primary zone and thermoplastic shear instability at γ_i (see Chapter 6). After combining (11.1) and (11.3), he obtained

$$\gamma_s = \frac{r^2 - 2r \sin \alpha + 1}{r \cos \alpha} \quad \text{and} \quad \gamma_{s\,\min} = \frac{2(1 - \sin \alpha)}{\cos \alpha} \quad \text{at } r = 1. \quad (11.18)$$

Then von Turkovich asserted that if $\gamma_{s\,\min} > \gamma_i$, continuous cutting cannot be stable; hence the chip should become discontinuous. But in his approach heat transfer was completely ignored, and no comparison with experiment was given.

11.4 Blanking and related processes

11.4.1 Blanking and mechanical models

Now we turn to blanking, cropping and related processes of metal removal. As described at the beginning of this chapter, the shear plane is roughly parallel to the punch velocity in blanking or cropping. Therefore the processes are mainly controlled by sliding. However, in some cases, as shown in Figs. 11.16(d. e), visible cracks appear ahead of the punch and the die. Hence, apart from the initial indentation, possible bending, friction and local tension, sliding and cracking are the two major factors in blanking. Figure 11.17 is a schematic representation of the punch-force–punch-displacement autographic diagram. Two aspects that predominate are the energy consumption at the peak load B and the steep drop of the load between points C and D. The peak load B usually appears after plastic deformation develops from A. The increasing part AB must be due to material hardening, while the peak load B has to be connected with a plastic load instability. From B to C plastic deformation continues, but now it is during a post-stability phase, and some microcracks or voids may be formed. Beyond point C, crack propagation occurs. For some mild steels point C is close to the peak load, and the onset of the load drop can effectively end the process.

Johnson and Slater (1967) gave a comprehensive survey of slow and fast blanking of metals at ambient and elevated temperatures, in which the effects of clearance, aspect ratio, punch speed, edges of the tool and die and temperature on the load–displacement diagram, the blank finish and so on have been elaborated. Here the emphasis is put on plastic instability.

The softening mechanism involved in plastic instability has been dealt with in several ways. Atkins (1980b) proposed that geometrical softening of the reduced section in blanking due to indentation can play a significant role.

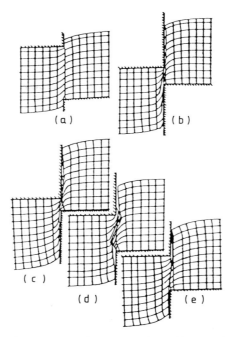

Fig. 11.16 Network diagrams showing progression of fracture in shearing metal bars (clearance nil): (a) lead 22% penetration; (b) lead 74% penetration; (c) aluminium 86% penetration; (d) copper 55% penetration; (e) brass 53% penetration. (After T. M. Chang and H. W. Swift, *J. Inst. Metals* **78**, 119–148 (1950).)

Then the punch load F can be expressed as

$$F = \pi d(t - \delta)\tau$$

$$= \pi d(t - \delta)\tau_0 \gamma^n$$

$$= \pi d(t - \delta)\tau_0 \left(\frac{u}{\delta}\right)^n, \tag{11.19}$$

if friction is neglected and power-law hardening is assumed, where d denotes the diameter of punch, t the thickness of the workpiece, u the punch displacement and δ the width of the shear band. The instability criterion $dF = 0$ leads to

$$u_{\text{peak}} = \frac{nt}{1 + n}. \tag{11.20}$$

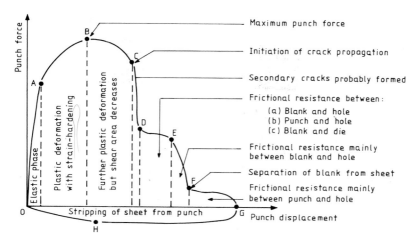

Fig. 11.17 Schematic representation of punch-force–punch-displacement autographic diagram. (After W. Johnson and R. A. C. Slater, in *Proc. CIRP–ASTME*, pp. 825–851 (1967).

This value has been compared with experimental values, and the agreement is good. Bai and Johnson (1982) suggested another softening mechanism—adiabatic shear instability. The model is mainly applicable to high-speed plugging without a die. They assumed that there was no macroscopic geometrical softening due to indentation, but a nonuniform distribution of shear strain in the workpiece adjacent to the punch edge. With deeper penetration, the shear strain tends to be concentrated, and finally leads to instability. This model has been used to explain the size-effect in plugging.

Later, Dodd and Atkins (1983a) and Dodd (1983) combined and compared the two mechanisms, and accordingly classified materials. Furthermore, they suggested a third softening mechanism—microvoids; then, from the work of Rice and Tracey (1969) described in §5.2.2, the term $t - \delta$ becomes $t - \delta - \delta_v$, which takes account of the void dimensions. The three mechanisms may be suitable for different materials and design of blanking processes. For example, adiabatic shear may occur for materials with poor heat conduction and low specific heat, and void mechanisms may well occur for materials with porosity. A high compression generated in the shearing zone, for instance by clamping, can prevent crack formation, which helps to achieve smooth shear surfaces and high-dimensional accuracy (Nakagawa and Maeda, 1970). A high compressive stress would clearly suppress void or crack nucleation, and this is a key feature in the production of smooth sheared surfaces.

11.4.2   Associated plastic flow and cracking

Ductile fracture has meant, until now, that the workpiece is fully or partially separated by gross plastic deformation, such as chips and work material or blank and work sheet. Generally speaking, ductile fracture may be regarded as occurring either without cracks, which we call rupture (see §1.4 and Fig. 1.1), or with cracking. The latter is much more common in ductile fractures.

In metal cutting and blanking, how is it possible to identify the crack? In metal cutting, if there were no cracks, how should the new surfaces of the chip and the workpiece, which are by no means a part of the original surface of the workpiece, be created? In blanking, does the penetration consist of plastic shear flow and crack extension, even though there is no visible crack? Perhaps a crack should be identified as a surface that is new topologically, i.e. it was the inner portion of material rather than a part of the original surface; mechanically the surfaces cannot withstand any tensile stresses, and metallurgically the surfaces appear to have certain fractographic features, unlike the appearance of the original surface or its deformed version. So, in this sense, the cut metal has been cracked. In blanking, on the other hand, the matter may be different. For some very ductile materials, the material has to undergo plastic shear instability, there is a peak load, but cracks may be suppressed, and then the steep load drop from point C in Fig. 11.17 would not appear until full penetration or say separation of the blank from the sheet. This is similar to rupture (Fig. 1.1). Perhaps the case of lead in Fig. 11.16(b) with 74% penetration approaches this state. Hence here plastic localization plays the predominant role in ductile fracture. However, as we have seen, ductile fracture usually consists of localization and macrocracking, which is the result of microscopic voids and shear bands. Therefore there is no one unique mechanism of ductile fracture, but many possible mechanisms; these may be combined and can be a mixture of macro- and microscopic mechanisms.

References

Abe, T., Nagaki, S. and Hayashi, T. (1983). In *Proc. 24th Polish Solid Mechanics Conf.*, pp. 1–6.

Allen, N. P., Hopkins, B. E. and McLennan, J. E. (1956). *Proc. R. Soc. Lond.* **A234**, 221–246.

Anand, L. and Spitzig, W. A. (1980). *J. Mech. Phys. Solids* **28**, 113–128.

Argon, A. S. and Im, J. (1975). *Met. Trans.* **6A**, 839–851.

Ashby, M. F. (1966). *Phil. Mag.* **14**, 1157–1178.

Ashby, M. F. (1981). *Prog. Mat. Sci.; Bruce Chalmers Anniversary Volume*, pp. 1–25.

Ashby, M. F., Gandhi, C. and Taplin, D. M. R. (1979). *Acta Metall.* **27**, 699–729.

Atkins, A. G. (1980a). The mechanics of ductile fracture in metalforming. Presented to Int. Cold Forging Group, Sept., Sentis, France.

Atkins, A. G. (1980b). *Int. J. Mech. Sci.* **22**, 215–231.

Atkins, A. G. (1981). Phil. Mag. **43**, 627–641.

Avitzur, B. (1968). Metal Forming: Processes and Analysis. McGraw-Hill, New York.

Avitzur, B. (1977). *Ann. Rev. Mat. Sci.* **7**, 261–300.

Azrin, M. and Backofen, W. A. (1970). *Met. Trans.* **1**, 2852–2865.

Backman, M. E. and Finnegan, S. A. (1973). In *Metallurgical Effects at High Strain Rates* (ed. R. W. Rohde, B. M. Butcher, J. R. Holland and C. H. Karnes), pp. 531–543. Met. Soc. AIME. Plenum Press, New York.

Backofen, W. A. (1972). *Deformation Processing*. Addison-Wesley, Reading, Massachusetts.

Bai, Y. (1980). In *Shock Waves and High-Strain-Rate Phenomena in Metals* (ed. M. A. Meyers and L. E. Murr), pp. 277–281. Plenum Press, New York.

Bai, Y. (1982). *J. Mech. Phys. Solids* **30**, 195–207.

Bai, Y. and Dodd, B. (1985). *Res. Mechanica* **13**, 227–241.

Bai, Y. and Johnson, W. (1982), *Met. Technol.* **9**, 182–190.

Bai, Y., Cheng, Z. M. and Yu, S. B. (1984). Institute of Mechanics Report, Chinese Academy of Sciences.

Baker, T. J. and Charles, J. A. (1972). *J. Iron Steel Inst.* **210**, 702–706.

Barba, M. J. (1880). *Mem. Soc. Ing. Civils* pt. 1, p. 682.

Barnby, J. T. (1967). *Acta. Metall.* **15**, 1213–1228.

Basinski, Z. S. (1957). *Proc. R. Soc. Lond.* **A240**, 229–242.

Bedford, A. J., Wingrove, A. L. and Thompson, K. R. L. (1974). *J. Aust. Inst. Metals* **19**, 61–73.

Beevers, C. J. and Honeycombe, R. W. K. (1959). In *Fracture: Proc. Swampscott Conf.* (ed. B. L. Averback, D. K. Felbeck, G. T. Hahn and D. A. Thomas), pp. 474–497. Wiley, New York.

Beevers, C. J. and Honeycombe, R. W. K. (1961). *Acta Metall.* **9**, 513–524.

Berg, C. A. (1962). In *Proc. 4th US Nat. Congr. on Applied Mechanics*, Vol. 2, pp. 885–892.

Blazynski, T. Z. (1976). *Metal Forming—Tool Profiles and Flow.* Macmillan, London.

Boothroyd, G. (1963). *Proc. Inst. Mech. Engrs* **177**, 789–802.

Boothroyd, G. (1981). *Fundamentals of Metal Machining and Machining Tools.* McGraw-Hill, New York.

Bourne, L. (1951). BISRA Report MW/A/23/51.

Brandes, M. (1970). In *Mechanical Behaviour of Materials under Pressure* (ed. H. Ll. D. Pugh), pp. 236–298, Elsevier, Amsterdam.

Bridgman, P. W. (1944). *Trans. ASM* **32**, 553–574.

Bridgman, P. W. (1952). *Studies in Large Plastic Flow and Fracture.* McGraw-Hill, New York.

Broek, D. (1971). A study on ductile fracture. NRL-TR-71021-U.

Broek, D. (1973). *Engng Frac. Mech.* **5**, 55–66.

Broek, D. (1982). *Elementary Engineering Fracture Mechanics.* Martinus Nijhoff, The Hague.

Brown, L. M. and Stobbs, W. M. (1976). *Phil. Mag.* **34**, 351–372.

Brownrigg, A., Spitzig, W. A., Richmond, O., Teinlinck, D. and Embury, J. D. (1983). *Acta Metall.* **31**, 1141–1150.

Brozzo, P., Deluca, B. and Rendina, R. (1972). A new method for the prediction of the formability limits of metal sheets. In *Proc. 7th Biennial Congr. of Int. Deep Drawing Research Group.*

Budiansky, B. (1984). In *Mechanics of Material Behaviour* (ed. G. J. Dvorak and R. T. Shield), pp. 15–29. Elsevier, Amsterdam.

Burdekin, F. M. and Stone, D. E. W. (1966). *J. Strain Anal.* **1**, 145–153.

Butcher, B. M., Barker, L. M., Munson, D. E. and Lundergan, C. D. (1964). *J of AIAA* **2**, 977–990.

Campbell, J. D. (1967). *J. Mech. Phys. Solids* **15**, 359–370.

Campbell, J. D. (1973). *Mat. Sci. Engng* **12**, 3–21.

Chang, T. M. and Swift, H. W. (1950). *J. Inst. Metals* **78**, 119–146.

Cheng, C. M. (1982). *Adv. Mech.* **12**, 133–140.

Childs, T. H. C., Richings, D. and Wilcox, A. B. (1972). *Int. J. Mech. Sci.* **14**, 359–375.

Chin, G. Y., Hosford, W. F. and Backofen, W. A. (1964a). *Trans. Met. Soc. AIME* **230**, 437–449.

Chin, G. Y., Hosford, W. F. and Backofen, W. A. (1964b). *Trans. Met. Soc. AIME* **230**, 1043–1048.

Christoffersen, J. and Hutchinson, J. W. (1979). *J. Mech. Phys. Solids* **27**, 465–487.

Clift, S. E. (1986). Identification of Defect Locations in Forged Products using the Finite Element Method. Ph.D. thesis, University of Birmingham.

Clift, S. E., Hartley, P., Sturgess, C. E. N. and Rowe, G. W. (1985). Fracture initiation in plane strain forging. In *Proc. 25th MTDR Conf.*, pp. 413–419.

Clifton, R. J. (1980). Report NMAB-356. National Materials Advisory Board (U.S.), 356, pp. 129–142.

Clyens, S. (1973). Experiments of dynamic material behaviour. M.Sc. thesis, University of Oxford.

Considère, A. (1885). *Ann. ponts et chaussees* ser. 6, **9**, 575–775.

Cook, N. H., Finnie, I. and Shaw, M. C. (1954). *Trans. ASME* **76**, 153–162.

Costin, L. S., Crisman, E. E., Hawley, R. H. and Duffy, J. (1979). In *Inst. Phys. Conf. Ser.* No. 47 (ed. J. Harding), pp. 90–100.

Cottrell, A. H. (1957). *Properties of Materials at High Rates of Strain*. Inst. Mech. Engrs, London.

Cottrell, A. H. (1959). In *Fracture: Proc. Swampscott Conf.* (ed. B. L. Averback, D. K. Felbeck, G. T. Hahn and D. A. Thomas), pp. 20–44. Wiley, New York.

Cottrell, A. H. (1964). *The Mechanical Properties of Matter*. Wiley, London.

Cottrell, A. H. and Bilby, B. A. (1949). *Proc. Phys. Soc.* **A62**, 49–69.

Culver, R. S. (1973). In *Metallurgical Effects at High Strain Rates* (Met. Soc. AIME) (ed. R. W. Rohde, B. M. Butcher, J. R. Holland and C. H. Karnes), pp. 519–530. Plenum Press, New York.

Curran, D. R., Seaman, L. and Shockey, D. A. (1977). *Phys. Today* **30**, 46–55.

Dao, K. C. and Shockey, D. A. (1979). *J. Appl. Phys.* **50**, 8244–8246.

Davidenkov, N. N. and Spiridonova, N. I. (1946). *Proc. ASTM* **46**, 1147–1158.

Davidson, T. E. and Ansell, G. S. (1969). *Trans. Met Soc. AIME* **245**, 2383–2390.

Davidson, T. E., Uy, J. C. and Lee, A. P. (1966). *Acta Metall.* **14**, 937–948.

Dieter, G. E. (1976). *Mechanical Metallurgy*, 2nd edn. McGraw Hill, New York.

Dodd, B. (1983). *Met. Technol.* **10**, 57–60.

Dodd, B. and Atkins, A. G. (1983a). *Acta Metall.* **31**, 9–15.

Dodd, B. and Atkins, A. G. (1983b). In *Developments in Drawing Metals*, pp. 87–89. Metals Society.

Dodd, B. and Boddington, P. (1980). *J. Mech. Working Techn.* **3**, 239–252.

Dodd, B. and Caddell, R. M. (1983). *Int. J. Mech. Sci.* **26**, 113–118.

Dodd, B. and Kudo, H. (1980). *Int. J. Mech. Sci.* **22**, 67–71.

Dodd, B. and Scivier, D. A. (1975). *Int. J. Mech. Sci.* **17**, 663–667.

Dodd, B., Stone, R. C. and Bai, Y. (1985). *Res. Mechanica* **13**, 265–273.

Doraivelu, S. M., Gapinathan, V. and Venkatesh, V. C. (1981). In *Shock Waves and High-Strain-Rate Phenomena* (ed. M. A. Meyers and L. E. Murr), pp. 263–275. Plenum Press, New York.

Drucker, D. C. (1949). *J. Appl. Phys.* **20**, 1013–1021.

Dugdale, D. S. (1960). *J. Mech. Phys. Solids* **8**, 100–108.

Duncombe, E. (1972). *Int. J. Mech. Sci.* **14**, 325–337.

Eary, D. F. and Reed, E. A. (1974). *Techniques of Pressworking Sheet Metal*. Prentice-Hall, London.

Edelson, B. I. and Baldwin, W. M. (1962). *Trans. ASM* **55**, 230–250.

Eleiche, A.-S. M. (1972). AAFML-TR-72-125.

El-Sabaie, M. G. and Mellor, P. B. (1972). *Int. J. Mech. Sci.* **14**, 535–556.

El-Sabaie, M. G. and Mellor, P. B. (1973). *Int. J. Mech. Sci.* **15**, 485–501.

Ernst, H. J. and Merchant, M. E. (1941). *Trans. ASM* **29**, 299–378.

Ertürk, T. (1979). In *Proc. Int. Conf. on the Mechanical Properties of Materials 3*, Vol. 2 (ed. K. J. Miller and R. F. Smith), pp. 653–662.

Ewalds, H. L. and Wanhill, R. J. H. (1984). *Fracture Mechanics*. Arnold/DUM, London/Delft. ✓

Fields, D. S. and Backofen, W. A. (1957). *Proc. ASTM* **57**, 1259–1272.

Fields, R. J., Weerasooriya, T. and Ashby, M. F. (1980). *Met. Trans.* **11A**, 333–347.

French, J. A. and Weinrich, P. F. (1975). *Met. Trans.* **6A**, 785–790.

Frost, H. J. and Ashby, M. F. (1982). *Deformation-Mechanism Maps—The Plasticity and Creep of Metals and Ceramics*. Pergamon Press, Oxford.

Furubayashi, T. and Kojima, M. (1982). In *Formability of Metallic Materials—2000 A.D.* (ed. J. R. Newby and B. A. Niemeier). ASTM STP 753, pp. 119–136.

Gandhi, C. and Ashby, M. F. (1979). *Acta Metall.* **27**, 1565–1602.

Gell, M. and Worthington, P. J. (1966). *Acta Metall.* **14**, 1265–1271.

Ghosh, A. K. (1978). In *Mechanics of Sheet Metal Forming* (ed. D. P. Koistinen and N.-M. Wang), pp. 287–312. Plenum Press, New York.

Gleiter, H. (1967). *Acta Metall.* **15**, 1213–1228.

Goodier, J. N. (1933). *J. Appl. Mech.* **1**, 39–44.

Goods, S. H. and Brown, L. M. (1979). *Acta Metall.* **27**, 1–15.

Goodwin, G. M. (1968). SAE Paper 680093.

Griffith, A. A. (1921). *Phil. Trans. R. Soc. Lond.* **A221**, 163–197.

Guillen-Preckler, A. (1967). *Ann. CIRP* **15**, 183–188.

Gurland, J. and Plateau, J. (1963). *Trans. ASM* **56**, 442–454.

Gurson, A. L. (1977). *Trans. ASME, H: J. Engng Mat. Technol.* **99**, 2–15.

Hahn, G. T. and Rosenfield, A. R. (1975). *Met. Trans.* **6A**, 653–670.

Hammerschmidt, M. and Kerye, H. (1981). In *Shock Waves and High-Strain-Rate Phenomena in Metals* (ed. M. A. Meyers and L. E. Murr), pp. 961–973. Plenum Press, New York.

Hancock, J. W. and Mackenzie, A. C. (1976). *J. Mech. Phys. Solids* **24**, 147–169.

Harding, J. (1981). In *Proc. 7th Int. Conf. on High Energy Rate Fabrication, University of Leeds, 1–9 Sept.*

Harris, J. N. (1983). *Mechanical Working of Metals—Theory and Practice*. Pergamon Press, Oxford.

Hart, E. W. (1967). *Acta Metall.* **15**, 351–355.

Hartman, K. H., Kunze, H. D. and Meyer, L. W. (1981). In *Shock Waves and High-Strain-Rate Phenomena in Metals* (ed. M. A. Meyers and L. E. Murr), pp. 325–337. Plenum Press, New York.

Hastings, W. F., Mathew, P. and Oxley, P. L. B. (1980). *Proc. R. Soc. Lond.* **A371**, 569–587.

Hecker, S. S. (1977). In *Formability: Analysis Modelling and Experimentation* (ed. S. S. Hecker, A. K. Ghosh and H. L. Gegel), pp. 150–182. Met. Soc. AIME.

Hecker, S. S. (1978). In *Application of Numerical Methods to Forming Processes* (ed. H. Armen and R. F. Jones), pp. 85–94. ASME AMD-28.

Hessenberg, W. C. F. and Bourne, L. (1949). BISRA Report MW/A/51/49.

Hill, R. (1948). *Proc. R. Soc. Lond.* **A193**, 281–297.

Hill, R. (1950). *Mathematical Theory of Plasticity*. Clarendon Press, Oxford.

Hill, R. (1952). *J. Mech. Phys. Solids* **1**, 19–30.

Hill, R. (1954). *J. Mech. Phys. Solids* **3**, 47–53.

Hill, R. (1957). *J. Mech. Phys. Solids* **5**, 153–161.

Hill, R. (1958). *J. Mech. Phys. Solids* **6**, 1–8.

Hill, R. (1962). *J. Mech. Phys. Solids* **10**, 1–16.

Hill, R. (1979). *Math. Proc. Camb. Phil. Soc.* **85**, 179–191.

Hill, R. and Hutchinson, J. W. (1975). *J. Mech. Phys. Solids* **23**, 239–264.

Hirsch, P. B. (1957). *J. Inst. Metals* **86**, 13–14.

Honeycombe, R. W. K. (1981). *Steels, Microstructure and Properties*. Arnold, London.

Hodge, P. G. (1970). *Continuum Mechanics*, p. 65. McGraw-Hill, New York.

Hornbogen, E. (1984). *Acta Metall.* **32**, 615–627.

Hosford, W. F. and Backofen, W. A. (1964). In *Fundamentals of Deformation*

Processing: Proc. 9th Sagamore Army Materials Research Conf. (ed. W. A. Backofen, J. J. Burke, L. F. Coffin, N. L. Reed and V. Weiss), pp. 253–292. Syracuse University Press, New York.

Hosford, W. F. and Caddell, R. M. (1983). *Metal Forming—Mechanics and Metallurgy.* Prentice-Hall, Englewood Cliffs, New Jersey.

Hull, D. (1975). *Introduction to Dislocations*, 2nd edn. Pergamon Press, Oxford.

Hutchinson, J. W. (1979). In *Proc. 8th US National Congr on Appl. Mechanics* (ed. R. E. Kelly), pp. 87–98. Western Periodicals Co., North Hollywood, California.

Hutchinson, J. W. (1981). In *Proc. Workshop on Flow Localization* (ed. R. J. Asaro and A. Needleman), pp. 46–47. Brown University Report MRLE-127.

Hutchinson, J. W. and Miles, J. P. (1974). *J. Mech. Phys. Solids* **22**, 61–71.

Hutchinson, J. W. and Neale, K. W. (1978). In *Mechanics of Sheet Metal Forming* (ed. D. P. Koistinen and N.-M. Wang), pp. 127–150. Plenum Press, New York.

Hutchinson, J. W. and Tvergaard, V. (1981). *Int. J. Solids and Structures* **17**, 451–470.

Irwin, C. J. (1972). *Metallographic Interpretation of Impacted Ogive Penetrators.* DREV-R-652/72, Canada.

Irwin, G. and Wells, A. A. (1965). *Met. Rev.* **10**, 223–270.

Ishigaki, H. (1978). In *Mechanics of Sheet Metal Forming* (ed. I. Koistinen and N.-M. Wang), pp. 315–338. Plenum Press, New York.

Iwata, K., Osakada, K. and Terasaka, Y. (1984). *Trans. ASME, H: J. Engng Mat. Technol.* **106**, 132–138.

Jain, S. C. and Kobayashi, S. (1970). In *Proc. 11th Int. Machine Tool Design Research Conf.* (ed. S. A. Tobias), pp. 1137–1154. Macmillan, London.

Jalinier, J.-M. (1981). Mise en forme et endommagement. Doctor of Engineering thesis, University of Metz.

Jenner, A. and Dodd, B. (1981). *J. Mech. Working Technol.* **5**, 31–43.

Jenner, A., Bai, Y. and Dodd, B. (1981). *J. Strain Anal.* **16**, 159–164.

Jennison, H. C. (1930). *Trans. AIMM* **89**, 121–139.

Johnson, W. (1971). *Appl. Mech. Rev.* **25**, 977–989.

Johnson, W. (1972a). *Impact Strength of Materials.* Arnold, London.

Johnson, W. (1972b). *Metallurgia and Metal Forming* **39**, 89–99.

Johnson, W. and Mamalis, A. G. (1977). In *Proc. 17th Int. Machine Tool Design Research Conf.* (ed. S. A. Tobias), pp. 607–621. Macmillan, London.

Johnson, W. and Mellor, P. B. (1983). *Engineering Plasticity.* Ellis Horwood, Chichester, England.

Johnson, W. and Slater, R. A. C. (1967). In *Proc. CIRP–ASTME*, pp. 825–851.

Johnson, W., Baraya, G. L. and Slater, R. A. C. (1964). *Int. J. Mech. Sci.* **6**, 409–414.

Johnston, W. G. and Gilman, J. J. (1959). *J. Appl. Phys.* **30**, 129–144.

Kármán, T. von (1911). *Z. Vereins deutscher Ing.* **55**, 1749–1757.

Keeler, S. P. (1965). *Sheet Metal Industries* **42**, 683–691.

Keeler, S. P. (1968). SAE Paper 680092.

Keeler, S. P. and Backofen, W. A. (1963). *Trans. ASM* **56**, 25–48.

Kelly, A. and Davies, G. J. (1965). *Met. Ref.* 1–144.

Kiessling, R. and Lange, N. (1966). *Non-Metallic Inclusions in Steel* (Part II). Iron and Steel Inst., London.

Kikuchi, M., Shiozawa, K. and Weertman, J. R. (1981). *Acta Metall.* **29**, 1747–1758.

Kivivuori, S. and Sulonen, M. (1978). *Ann. CIRP* **27**, 141–145.

Knott, J. F. (1973). *Fundamentals of Fracture Mechanics.* Butterworths, London.

Kobayashi, H. and Dodd, B. (1985). Dept Engng, Univ. Reading, Report 102/85.

Kobayashi, S. (1970). *Trans. ASME, B: J. Engng Ind.* **92**, 391–399.

Kopalinsky, E. M. and Oxley, P. L. B. (1984). In *Mechanical Properties at High Rates of Strain* (ed. J. Harding), pp. 389–396. Conf. Ser. No. 70, Inst. Phys., London.

Korhonen, A. S. (1978). *Trans. ASME*, H: *J. Engng Mat. Technol.* **100**, 303–309.

Korhonen, A. S. (1981). Necking, fracture and localization of plastic flow in metals. Doctor of Technology thesis, Helsinki University of Technology.

Kudo, H. (1960). *Int. J. Mech. Sci.* **2**, 102–127.

Kudo, H. (1965). *Int. J. Mech. Sci.* **7**, 43–55.

Kudo, H. and Aoi, K. (1967). *J. Japan Soc. Tech. Plasticity* **8**, 17–27.

Kudo, H. and Tsubouchi, M. (1971). *Ann. CIRP* **19**, 225–230.

Kudo, H., Sato, K. and Aoi, K. (1968). *Ann. CIRP* **16**, 309–318.

Kudo, H., Nagahama, T. and Yoshida, K. (1972). *Int. J. Mech. Sci.* **14**, 339–342.

Kuhn, H. A. (1978). In *Advances in Deformation Processing: Proc. 21st Sagamore Army Materials Res. Conf.* (ed. J. J. Burke and V. Weiss), pp. 159–186.

Landau, L. D. and Lifshitz, E. M. (1959). *Fluid Mechanics.* Pergamon Press, Oxford.

Larsen-Basse, J. and Oxley, P. L. B. (1973). In *Proc. 13th Machine Tool Design Research Conf.* (ed. S. A. Tobias and F. Koenigsberger), pp. 209–216. MacMillan, London.

Latham, D. J. (1963). The effect of stress system on the workability of metals. Ph.D. thesis, University of Birmingham.

Latham, D. J. and Cockcroft, M. G. (1966). NEL Report No. 216.

Lee, D. (1984). *Trans ASME*, H: *J. Engng Mat. Technol.* **109**, 9–15.

Lee, E. H. and Shaffer, B. W. (1951). *J. Appl. Mech.* **73**, 405–413.

Lee, P. W. (1972). Fracture in cold forming of metals—a criterion and a model. Ph.D. thesis, Drexel University.

Lee, P. W. and Kuhn, H. A. (1973). *Met. Trans.* **4**, 969–974.

Lemaire, J. C. and Backofen, W. A. (1972). *Met. Trans.* **3**, 477–482.

LeRoy, G., Embury, J. D., Edwards, G. and Ashby, M. F. (1981). *Acta Metall.* **29**, 1509–1522.

Leslie, W. C. (1982). *Howe Memorial Lecture, Iron and Steel Soc. of AIME*, pp. 3–50.

Lin, I.-H., Hirth, J. P. and Hart, E. W. (1981). *Acta Metall.* **29**, 819–827.

Lindholm, U. S. (1974). In *Mechanical Properties at High Rates of Strain* (ed. J. Harding). Conf. Ser. No. 21, Inst. Phys., London.

Lindholm, U. S. and Bessey, R. L. (1969). AAFML-TR-69-119.

Lindholm, U. S., Nagy, A., Johnson, G. R. and Hoegfeldt, J. M. (1980). *Trans. ASME*, H: *J. Engng Mat. Technol.* **102**, 376–381.

Loewen, E. G. and Shaw, M. C. (1954). *Trans. ASME* **76**, 217–231.

Low, J. R., Van Stone, R. H. and Merchant, R. H. (1972). NASA Tech. Report 2, NGR-38-087-003, Carnegie-Mellon Univ.

Ludwik, P. (1927). *Z. Vereins deutscher Ing.* **71**, 1532–1538.

McClintock, F. A. (1968). *J. Appl. Mech.* **90**, 363–371.

McClintock, F. A., Kaplan, S. M. and Berg, C. A. (1966). *Int. J. Fract. Mech.* **2**, 614–627.

MacGregor, C. W. (1937). AIME Tech. Pub. 805.

MacGregor, C. W. and Fisher, J. C. (1946). *J. Appl. Mech.* **13**, A11–A16.

Machida, T. and Nakagawa, T. (1978). *J. Prod. Res.* **90**, 61–64.

McLean, D. (1962). *Mechanical Properties of Metals.* Wiley, London.

Malvern, L. E. (1969). *Introduction to the Mechanics of a Continuous Medium.* Prentice-Hall, Englewood Cliffs, New Jersey.

Marciniak, Z. (1965). *Arch. Mech. Stoswanej* **4**, 577–592.

Marciniak, Z. and Kuczynski, K. (1967). *Int. J. Mech. Sci.* **9**, 609–620.

Martin, J. W. (1968). *Precipitation Hardening.* Pergamon Press, Oxford.

Martin, J. W. (1980). Micromechanisms in Particle-hardened Alloys, Cambridge Univ. Press, Cambridge.

Meester, B. de and Tozawa, Y. (1979). *Ann. CIRP* **28**, 577–580.

Mellor, P. B. (1982). In *Mechanics of Materials: The Rodney Hill 60th Anniversary Volume* (ed. H. G. Hopkins and M. J. Sewell), pp. 383–415. Pergamon Press, Oxford.

Miller, L. E. and Smith, G. C. (1970). *J. Iron and Steel Inst.* **208**, 998–1005.

Mises, R. von (1928). *Z. angew. Math. Mech.* **8**, 161–185.

Moss, G. L. (1981). In *Shock Waves and High-Strain-Rate Phenomena in Metals* (ed. M. A. Meyers and L. E. Murr), pp. 299–312. Plenum Press, New York.

Nadai, A. (1931). *Plasticity.* McGraw-Hill, New York.

Nadai, A. (1950). *Theory of Flow and Fracture of Solids.* McGraw-Hill, New York.

Nakagawa, T. and Maeda, T. (1970). In *Proc. 4th Int. Meeting on Cold Forging*, pp. 351–376.

Nakamura, K. and Nakagawa, T. (1982). Preprint 223, Spring Conf. on Plastic Working (in Japanese).

Nakamura, K. and Nakagawa, T. (1983). Preprint 427, Spring Conf. on Plastic Working (in Japanese).

Nakamura, K. and Nakagawa, T. (1984). In *Proc. of 1st Conf. on Technology of Plasticity, Advanced Technology of plasticity*, pp. 775–788. J. Soc. Technology of Plasticity and J. Soc. of Precision Engng.

Neale, K. W. (1981). *S.M. Arch.* **6**, 79–128.

Needleman, A. (1972). *J. Mech. Phys. Solids* **20**, 111–127.

Needleman, A. and Tvergaard, V. (1977). *J. Mech. Phys. Solids* **25**, 159–183.

Negroni, F., Kobayashi, S. and Thomsen, E. G. (1968). *Trans ASME, B: J. Engng Ind.* **90**, 387–392.

Nichols, F. A. (1980). *Acta Metall.* **28**, 663–674.

Ng, P., Chakrabarty, J. and Mellor, P. B. (1976). In *Proc. 15th Int. Machine Tool Design Research Conf.* (ed. S. A. Tobias and F. Koenigsberger), pp. 579–586. Macmillan, London.

Og, S. I. and Kobayashi, S. (1976a). AAFML-TR-76-61.

Oh, S. I. and Kobayashi, S. (1976b). *Trans ASME, B: J. Engng Ind.* **98**, 800–806.

Oh, S. I., Chen, C. C. and Kobayashi, S. (1979). *Trans ASME, B: J. Engng Ind.* **101**, 36–44.

Olson, G. B., Mescall, J. F. and Azrin, M. (1981). In *Shock Waves and High-Strain-Rate Phenomena in Metals* (ed. M. A. Meyers and L. E. Murr), pp. 221–248. Plenum Press, New York.

Orowan, E. (1947). In *Proc. Symp. on Internal Stresses*, p. 451. Inst. Metals, London.

Orowan, E. (1950). *Fatigue and Fracture of Metals: MIT Symp.* Wiley, New York.

Orowan, E. (1955). *Welding J.* **34**, 1575–1605.

Osakada, K., Limb, M. and Mellor, P. B. (1973). *Inst. J. Mech. Sci.* **15**, 291–307.

Osakada, K., Watadani, A. and Sekiguchi, H. (1977). *Bull. JSME* **20**, 1557–1565.

Osakada, K., Watadani, A. and Sekiguchi, H. (1978). *Bull. JSME* **21**, 1236–1243.

Osakada, K., Koshijima, J. and Sekiguchi, H. (1981). *Bull. JSME* **24**, 534–539.

Oyane, M. (1972). *Bull. JSME* **15**, 1507–1513.

Painter, M. J. and Pearce, R. (1974). *J. Phys. D: Appl. Phys.* **7**, 992–1002.

Parmar, A. and Mellor, P. B. (1978). *Int. J. Mech. Sci.* **20**, 385–391.

Palmer, I. G., Smith, G. C. and Warda, R. D. (1966). In *Proc. Conf. on Physical Basis of Yield and Fracture* (ed. A. C. Strickland), pp. 53–59. Inst. Phys./Phys. Soc., London.

Pearson, C. E. (1953). *The Extrusion of Metals.* Chapman and Hall, London.

Pickering, F. B. (1978). *Physical Metallurgy and Design of Steels*. Applied Science Publishers, London.

Peirce, D., Asaro, R. J. and Needleman, A. (1982). *Acta Metall.* **30**, 1087–1119.

Piispanen, V. (1937). *Eripaines Teknillisesla Aikakauslehdesla* **27**, 315–322.

Porter, D. A., Easterling, E. A. and Smith, G. D. W. (1978). *Acta Metall.* **26**, 1405–1422.

Price, R. J. and Kelly, A. (1964). *Acta Metall.* **12**, 979–992.

Pugh, H. Ll. D. (1958). In *Proc. Conf. on Technology of Engineering Manufacture*, pp. 237–254.

Pugh, H. Ll. D. (1964). *ASTM Special Technical Publication 374*.

Pugh, H. Ll. D. (1965). *Bulleid Memorial Lectures*, Vol. IIIB. University of Nottingham.

Pugh, H. Ll. D. and Green, D. (1964/5). *Proc. Inst. Mech. Engrs* **179**, 415–437.

Puttick, K. E. (1959). *Phil. Mag.* **4**, 964–969.

Recht, R. F. (1964). *J. Appl. Mech.* **31**, 189–193.

Rice, J. R. (1968). *J. Appl. Mech.* **35**, 379–386.

Rice, J. R. (1977). In *Theoretical and Applied Mechanics* (ed. W. T. Koiter), pp. 207–220. North-Holland, Amsterdam.

Rice, J. R. and Tracey, D. M. (1969). *J. Mech. Phys. Solids* **17**, 201–217.

Ritchie, R. O. (1983). *Trans. ASME*, H. *J. Engng Mat. Technol.* **105**, 1–7.

Rogers, H. C. (1960). *Trans. Met. Soc. AIME* **218**, 498–506.

Rogers, H. C. (1968). In *Proc. ASM Ductility Seminar*, pp. 31–61. ASM, Ohio.

Rogers, H. C. (1971). In *Metal Forming—Interrelation Between Theory and Practice* (ed. A. L. Hoffmanner), pp. 453–459. Plenum Press, New York.

Rogers, H. C. (1974). Adiabatic shearing—a review. Drexel Univ. Report.

Rogers, H. C. (1979). *Ann. Rev. Mat. Sci.* **9**, 283–311.

Rogers, H. C. and Coffin, L. F. (1971). *Int. J. Mech. Sci.* **13**, 141–155.

Rosenfield, A. R. (1968). *Met. Rev.* **13** (No. 121), 29–40.

Rowe, G. W. (1977). *Principles of Industrial Metalworking Processes*. Arnold, London.

Rozovsky, E., Hahn, W. C. and Avitzur, B. (1973). *Met. Trans.* **4**, 927–930.

Rudnicki, J. W. and Rice, J. R. (1975). *J. Mech. Phys. Solids* **23**, 371–394.

Sachs, G. (1951). *Principles and Methods of Sheet-Metal Fabricating*, pp. 9–14. Reinhold, New York.

Sachs, G. (1954). *Fundamentals of the Working of Metals*. Pergamon Press, Oxford.

Saje, M., Pan, J. and Needleman, A. (1982). *Int. J. Fracture* **19**, 163–183.

Schmid, E. and Boas, W. (1950). *Plasticity of Crystals*. F. A. Hughes, London.

Schmidt, A. O. and Roubik, J. R. (1949). *Trans ASME* **71**, 245–248.

Schmitt, J. H. and Jalinier, J. M. (1982). *Acta Metall.* **30**, 1789–1798.

Seaman, L., Curran, D. R. and Shockey, D. A. (1976). *J. Appl. Phys.* **47**, 4814–4826.

Sekiguchi, H. and Osakada, K. (1983). *Ann. CIRP* **32**, 181–185.

Semiatin, S. L. and Jonas, J. J. (1984). *Formability and Workability*. ASM, Air Force Mat. Lab., Ohio.

Sharma, C. S., Rice, W. B. and Salmon, R. (1971). *Ann. CIRP* **19**, 545–549.

Shaw, M. C. (1984). *Metal Cutting Principles*. Clarendon Press, Oxford.

Slater, R. A. C. (1965/6). *Proc. Manchester Assn Engrs*, pp. 1–45.

Spretnak, J. W. (1974). AFML-TR-74-160.

Stakgold, J. (1971). *SIAM Rev.* **13**, 289–332.

Steif, P. S., Spaepen, F. and Hutchinson, J. W. (1982). *Acta Metall.* **30**, 447–456.

Stevenson, M. and Oxley, P. L. B. (1970). *Proc. Inst. Mech. Engrs* **185**, 55–71.

Stören, S. and Rice, J. R. (1975). *J. Mech. Phys. Solids* **23**, 421–441.

Suzuki, H., Hashizume, S., Yabuki, Y., Ichihara, Y., Nakajima, S., Kenmochi, K. (1968). *Rep. Inst. Ind. Sci., Univ. Tokyo* **18** (No. 3), 1–102.

Swift, H. W. (1952). *J. Mech. Phys. Solids* **1**, 1–18.

Tegart, W. J. M. (1966). *Elements of Mechanical Metallurgy.* Macmillan, London.

Thomason, P. F. (1968). *J. Inst. Metals* **96**, 360–365.

Thomason, P. F. (1981). *Acta Metall.* **29**, 763–777.

Trent, E. M. (1977). *Metal Cutting.* Butterworths, London.

Turkovich, B. F. von (1982). *On the Art of Cutting Metals—75 Years Later: A Tribute to F. W. Taylor. PED.* **7**, 85–98.

Tvergaard, V. (1981). *Int. J. Fracture* **17**, 389–407.

Tvergaard, V., Needleman, A. and Lo, K. K. (1981). *J. Mech. Phys. Solids* **29**, 115–142.

Walker, T. J. and Shaw, M. C. (1969). In *Proc. 10th Int. Machine Tool Design Research Conf.* (ed. S. A. Tobias and F. Koenigsberger), pp. 241–252. Pergamon Press, Oxford.

Watts, A. B. and Ford, H. (1952). *Proc. Inst. Mech. Engrs* **18**, 448–453.

Weiner, J. H. (1955). *Trans. ASME* **77**, 1331–1341.

Wells, A. A. (1962). In *Proc. Crack Propagation Symp.*, M13, Cranfield, Vol. 1, pp. 210–230.

Wessell, E. G. (1960). *Practical Fracture Mechanics for Structural Steel.* Paper H. UKAEA/Chapman and Hall, London.

Wilsdorf, H. G. F. (1983). *Mat. Sci. Engng* **59**, 1–39.

Wilson, D. V. (1971). In *Proc. Conf. on Effects of Second Phase Particles on the Mechanical Properties of Steel*, pp. 28–36. The Iron and Steel Inst.

Wingrove, A. L. (1973). *Met. Trans.* **1**, 219–224.

Woodthorpe, J. and Pearce, R. (1970). *Int. J. Mech. Sci.* **12**, 341–347.

Wray, P. J. (1969). *J. Appl. Phys.* **40**, 4018–4029.

Yamamoto, H. (1978). *Int. J. Fracture* **14**, 347–365.

Zener, C. (1948). *Fracturing of Metals*, pp. 3–31. ASM, Ohio.

Zener, C. and Hollomon, J. H. (1944). *J. Appl. Phys.* **15**, 22–32.

Author Index

Subject Index